T0265671

HOW TO GROW THE FLOWERS

For Jem, Bruno and Cass

Photography by Aloha Shaw

Marianne Mogendorff
& Camila Romain

HOW TO GROW THE FLOWERS

A sustainable approach to enjoying flowers through the seasons

PAVILION

Pavilion
An imprint of HarperCollins*Publishers* Ltd
1 London Bridge Street
London SE1 9GF

www.harpercollins.co.uk

HarperCollins*Publishers*
Macken House,
39/40 Mayor Street Upper,
Dublin 1
D01 C9W8
Ireland

10 9 8 7 6 5 4 3

First published in Great Britain by
Pavilion, an imprint of HarperCollins*Publishers* Ltd 2022

A catalogue record for this book is available from the British Library.

ISBN 9781911682011

This book contains FSC™ certified paper and other controlled sources to ensure responsible forest management.

For more information visit: www.harpercollins.co.uk/green

Reproduction by Rival Colour Ltd, UK
Printed and bound in Latvia by PNB Print

Photography: Aloha Shaw
Design Manager & Art Direction: Alice Kennedy-Owen
Project Editor: Sophie Allen

CONTENTS

FOREWORD

Growers and gardeners who are career changers are uniquely evangelical. I know this to be true because I am one, and it's one of the many things that I have in common with Marianne and Camila. We take nothing about working with plants for granted; every part of the process is a revelation. This wonder is what makes this book from the force behind Wolves Lane Flower Company an exquisite and generous offering.

I had the privilege of working in a greenhouse adjacent to Marianne and Camila in North London for one glorious season, and it was there that I was able to witness the sheer volume of love, graft and dedication they both put into farming flowers with regenerative, nature-centric principles at the heart of their work. As a food grower, my focus is most often on taste and, for a time, I failed to appreciate the significance of growing the kinds of plants that confer beauty. Marianne and Camila changed that for me. Carrying home a backpack filled with their sunflowers and offering a stem to each person I saw smile on seeing them is how I came to see the importance of what they do.

Under the shadow of unfolding climate catastrophe and biodiversity loss, I find myself filled with questions. How might we practise joy? How can we express ourselves artistically? How can we cultivate splendour on a planet that is being systematically dishonoured? Well, we can grow nectar-rich blooms that are alluring to both the human and the more-than-human and do so in a manner that is respectful of the natural systems that uphold us. We can encourage flowering plants to thrive so they blossom into reminders of what we are being called to protect. We can grow the flowers that adorn our celebrations and, when they wither, compost them to nourish the soil in which we might grow the plants of our collective future.

Marianne and Camila operate at the intersection of so much of what we need in this moment. They embody a practice of nurturing the soil and growing that is whole-hearted and earthy, regenerative and generous, and offers up the bountiful beauty and abundance that we so require as human beings. That is what their gorgeous book will help you to do too. It will show you how to grow flowers seasonally, sustainably and successfully. And it will encourage you to grow the types of plants that will gladden your heart, call the butterflies and bees to share your garden and be tied into scented posies to offer to those you love.

If you take all the advice and lessons in this book to heart, your flower-growing adventures will undoubtedly be magnificent.

Claire Ratinon, organic food grower and writer

WOLVES·LANE
FLOWER·COMPANY

INTRODUCTION

How to Grow the Flowers is a compilation of lessons learned in seasonal and sustainable flower growing. Our hope is that it will enable and empower flower enthusiasts to grow and enjoy flowers at home, finding beauty throughout the entire year. It is a manifesto, of sorts, for change in the floral industry, and a nod to our floral hero Constance Spry. Spry's *How to Do the Flowers* was written in 1952, an iconic pocket-sized book that already says so much of what we are passionate about: seasonality. Spry inspired generations of florists and flower enthusiasts to use and spot the best ingredients available, those that reflect a unique moment in time, and to work with the natural beauty of flowers – their curves, imperfections and growing habits. Her aim was always to reflect nature back at herself from the vase. Embracing the seasons is the first step to a more sustainable relationship with flowers and we hope our book will inspire you to appreciate the fleeting beauty of truly seasonal flowers.

If we can, you can

This book isn't the musings of two veteran flower farmers, growing on acres of land. There are a host of trailblazing (mostly) women who have been growing flowers for generations before us and have written all our well-thumbed favourite flower-growing bibles. However, as relative newcomers, just four years in, we know how paralysing it can be to translate the information and images in those books into even the smallest patch of productive growing space. Sometimes the sheer abundance of beauty and opportunity in those pages are enough to intimidate even the most confident gardener. We are just two thirty-somethings, career changers, without formal training or any qualifications in horticulture.

We're urban growers, growing on a tiny scrap of land sandwiched between terraced houses and a cemetery on Wolves Lane in North London. We hope that our warts-and-all approach to sharing our flower-growing journey, full of epic failures and head-in-hands mistakes, will boost the confidence of flower lovers whatever the space you have to grow in: from south-facing 30m/100ft gardens to postage-stamp sized patios. Really anyone can grow flowers.

Seasonal = sustainable

Back in April 2017 when we sowed our first cornflower seeds, we set ourselves two imperatives for our tiny enterprise: the flowers had to be both seasonal and sustainable. You'll see that there are instances of abundance in this book when we've succumbed to a certain amount of pictorial flowery exhibitionism. We want to take you on a tour of the seasons, not what flowers can be found at the flower market or the florist's shop but what we can grow in our climate throughout the year. Within this context, in the northern hemisphere it means that there will inevitably be a fallow period when fresh flowers are scarce and we have to wait for the new season to begin. At Wolves Lane we're well versed in the joy of anticipation. When you plant some tulips in November and wait until April the following year to see them bloom you fully appreciate the happiness and profound connection with nature that flowers offer.

We don't earn even half as much money as we once did, but our relationship with the earth and the alchemical process of growing plants has been galvanising, grounding and satisfying. Since we set up Wolves Lane Flower Company we've

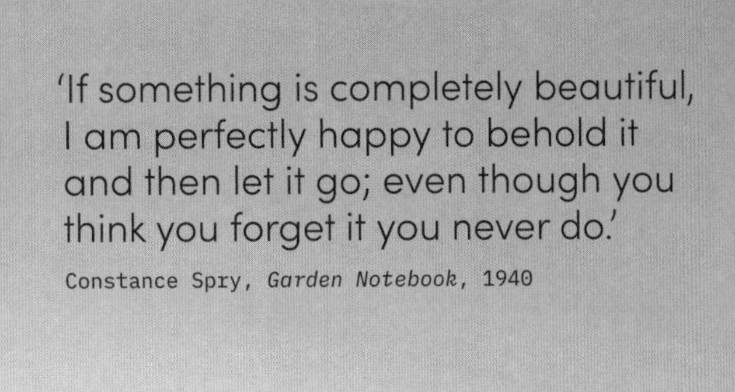

'If something is completely beautiful, I am perfectly happy to behold it and then let it go; even though you think you forget it you never do.'

Constance Spry, *Garden Notebook*, 1940

'This joyful experience is not limited to those who can grow or buy rare and expensive flowers, but is for everyone, school-child or student, town or country-woman, for everyone who loves a beautiful thing and will take a little trouble.'

Constance Spry, *How to Do the Flowers*, 1952

grown thousands of flowers and produced three baby boys between us. While having children is in some ways a huge life distraction, they have given us enormous focus and drive in our environmental efforts. There's nothing more terrifying than staring down the barrel of the climate crisis through the eyes of our children, and their children. It's with this in mind that we've written about growing flowers and about those two imperatives: seasonality and sustainability.

Helen Lewis, the Creative Director from Pavilion Books, who commissioned this book, first came across us while on the hunt for 'Belle Epoque' tulips for a photo shoot. What started off as a conversation about spring bulbs ended with the suggestion that we might like to write a book on growing flowers. At the time, we had only been growing for two years, so we howled with laughter and baulked at the idea that anyone would read anything that we had written on the subject. We were already members of Flowers from the Farm, the UK's flower-farming network for small-scale growers, so we were acquainted with some of the female powerhouses of the industry and shuddered to think of their reaction to two urban upstarts writing anything supposedly new about flower farming. Helen Lewis waited it out and persisted, albeit gently, for a couple of years. Fast forward to March 2020 and the world closed down, Covid struck and everyday life was put on hold. If we'd been sceptical about writing a book on flower growing pre-Covid, the pandemic changed all of that. Our tiny business was inundated with requests for people wanting to volunteer, career changers wanting to retrain, florists wanting to source British flowers for the first time in their lengthy careers, Londoners wanting to send flowers to loved ones; it seemed that the void left by human contact was being filled with flowers.

So it felt like the right moment to harness that collective need for flowers into a book that we felt was missing. While *How to Grow the Flowers* is about growing seasonally at home, we also wanted it to spill the beans on an industry that falls short when it comes to sustainability. An industry that regularly exploits the idea of a natural ecological product without any clear commitment to safeguarding environmental standards, habitats and the welfare of the people growing the flowers.

We hope that the information that we've shared will encourage flower lovers and consumers to start asking the right questions about the provenance of their flowers, to stop thinking about flowers as just another commodity but as the product of a fragile and vital ecosystem, one that we are all part of and all responsible for protecting.

Some context

The global cut-flower industry is worth approximately $29.2 billion and projected to grow to $41.1 billion by 2027. That's a lot of flowers zipping around the globe, many of which are grown on flower farms close to the equator. The UK's flower consumption makes up a small part of that gargantuan figure with an industry worth £1.4 billion, still not a figure to be sneezed at. However, less than 15% of those flowers are grown on British soil and mostly by big growers working under acres of glass in Lincolnshire and Cornwall or over hectares and hectares of farmland.

A UK-grown bouquet is responsible for roughly 10% of the carbon emissions of its Kenyan or Dutch equivalent. As growers we know exactly what additives and amendments we include in our growing and harvesting processes – basically compost, microbial teas and white vinegar to clean our buckets. While pesticides, herbicides and fungicides applied to our food are regulated and scrutinised (although arguably not enough), the cut-flower trade is more like the wild west. The range of chemicals used in industrial farming is frankly terrifying – chemicals which are used to treat seeds, sprayed liberally throughout the growing period and then again in the post-harvest conditioning process.

Over the last fifty years, industrial flower production and consumption have peaked with very little awareness of the costs to the environment. We hope this book convinces you to swing the dial back. This book isn't about finger pointing or bad mouthing industrial farming. It is about informing flower lovers about what it takes to grow flowers – both abroad and at home – to reconnect us with flowers as living things inextricably linked to our health and happiness.

Just begin

Perhaps it's trite to say if we can do it, so can you, but the truth is that nature does nearly all of the

The unheated glasshouse at WLFC.

hard work for us. Encased in those little brown seeds is a complex genetic code that tells the plant exactly what conditions it needs to grow and thrive. The microorganisms in healthy soil provide the ideal growing conditions for that to happen. You become a conduit between earth and seed, intervening in a sort of botanical destiny for the plants; really they already know what they're doing.

This is not an expert gardening manual, nor is it a glossy coffee-table edition of irresistible floral designs. It is a testament to our passions: growing flowers and protecting the future of our planet. We hope that flower lovers and people new to growing will derive useful lessons from this book. Growing isn't about getting it right, it's about sticking close to the seasons and getting started. The cyclical nature of gardening means that there is no beginning, and there is always something you could have done six months before to set you up in a better position today. That's OK. The best thing to do is just begin.

We've structured this book into the four seasons, starting with autumn, as really this is the beginning of any grower's calendar. Within each of the seasons there are four subsections, Soil, Seed, Tend and Harvest, to give you the rundown on the essential jobs or things to know at any given time of year. In the UK our seasons are still reasonably pronounced and distinct – although this already feels like it's changing – but we hope that flower lovers and people new to growing across the planet will find useful lessons in this book. We've included a floral project for each season and some useful resources at the back, including a glossary to help debunk some of that, at times, mystifying 'gardener speak'. Occasionally we've included some wise and pertinent words from floral designer extraordinaire, Constance Spry. Dip in and out or follow through the year to get in step with the cycle. Just get growing!

Flower projects, mechanics and how-tos

Among the images of floral arrangements and seasonal bunches in this book you will find four seasonal flower projects. While we don't consider ourselves elite floral designers we do love to be creative with the flowers that we grow, and we encourage you to get creative too. We've provided a simple how-to guide for each project so that you can try them at home, and we have listed all the mechanics and sundries that you will need. All sorts of different methods and techniques can be used to achieve the same look, so if you're not confident in floristry, what we would say is just try experimenting a little and you'll soon find that it's mainly just a load of chicken wire and moss!

Sphagnum moss

The 30 species of moss that fall under the umbrella name of sphagnum moss are protected in the UK because they play a vital role in creating peat bogs, which are important habitats and carbon sinks. It's very important that you source your moss sustainably. Sphagnum moss doesn't look like the uniform bright green moss you'll find at a floristry wholesaler. It will be a mix of spongy textures and colours ranging from pink and orange to brown, so don't be surprised when you open the bag.

Sphagnum moss can hold eight times its weight in water so it's hugely valuable as a floristry sundry, but please make sure to use it again and again. If you forget to take your mechanics apart and your moss dries out, simply tip it back into the bag it came in and give it a generous watering, then leave it a cool, dark corner for a week and the moss will rehydrate.

A note on plastics

Using plastic in gardening is almost unavoidable. We do our best to keep our plastic consumption in check by repairing our pea and bean netting and weed-suppressing membrane if they become damaged. We avoid single-use plastic, and reuse everything from plant labels to our recyclable pots. We invest in sturdier plastic seed trays so that we can use them over multiple seasons. We don't use any cellophane when wrapping or transporting our flowers for clients.

'The object of a flower-holder is to keep the stems firmly in any position that may be required. I am yet to find anything which achieves this better than crumpled up large-mesh thin wire netting.'

Constance Spry, *Winter and Spring Flowers*, 1951

SEASONAL BUNCH

If there's one thing that differentiates growers from gardeners it's that all garden plants are fair game for the vase. Nothing is safe. Once you switch on your cutting radar, ideas for ingredients to trial are everywhere. The head gardeners and garden proud we know are rarely fans of letting us scissor-wielding florists loose among their herbaceous borders. They have painstakingly created a composition of colour and texture and this fragile, fleeting beauty can feel like something to preserve as long as possible, not snip at its peak. While we're all for an abundant garden, there is no harm in some light editing in your backyard to bring the outside in.

No floristry qualifications are needed; we're not suggesting a painstaking exercise in composition and colour but rather a quick snip to capture a moment in time. To truly appreciate the ebb and flow of a season and how each year of growing is unique, nothing beats getting out and cutting a handful. On each day we photographed Wolves Lane for this book we cut a bunch of what caught our eye. What was at its best just then. Due to print deadlines, winter is a little under-represented here, but don't let that put you off. There is *always* something to snip if you look hard enough.

'Agreeable as it may be to rush off to the shops and buy exotic and out of season flowers something is lost if the materials natural to the season are overlooked.'

Constance Spry, *Winter and Spring Flowers*, 1951

8th September

13th October

19th October

5th November

1st December

26th March

5th April

5th May

2nd June

17th June

8th July

13th August

AU TU MN

AUTUMN

WHY DID WE START GROWING FLOWERS?

If you've picked up this book then we hope that we won't need to work too hard to convince you to sow some seeds or take a cutting or two. We weren't so much convinced as propelled to start growing our own flowers. Ten years of working as producers in the fashion and theatre industries respectively had given us the enormous privilege of working with fantastic creatives on out-of-this-world projects. Weren't we lucky? But for anyone who doesn't know what a producer does, they look at a computer screen for most of the day, appease stressed creatives and juggle conference calls and site visits. A decade of Excel spreadsheets is enough to make anyone reconsider if they really like what they're doing and if they want to do it for another ten years. We realised how unhappy we were in our 'fabulous' laptop-bound London jobs. It sounds like a cliché, but we longed for an existence, let alone a job, with more connection with nature, with life out of doors, in step with the seasons. We lived gardenless throughout our twenties, always spying on what the downstairs neighbours were growing, making trips back home to luxuriate in the familiar hug of our parents' back gardens.

The trigger was Marianne's wedding in 2015 when she cobbled together a band of friends and neighbours to put together the flowers for her West Country wedding. All the flowers came from the gardens of family friends who offered whatever they had a glut of at that particular moment. Delphiniums, sweet Williams and cornflowers arrived by the armload. All the rest came from Organic Blooms, trailblazing organic flower farmers based just ten minutes away from Marianne's childhood home. After seeing what the British flower season could offer, we were hooked. We joked about a life of soil, working together in the future, living the good life, probably eating a lot of soup but surrounded by flowers. Admittedly we had both always loved flowers and gardens, but the floristry industry had always presented us with more problems than we could solve. We worried about the carbon footprint of imported stems, the use of damaging pesticides and fertilisers and the unknown working standards and welfare of the people, mostly women, at the bottom of the flower chain, working on these far-away flower farms. Solving these problems will be the work of generations of florists, flower lovers and farmers. Our only option was to grow them ourselves.

How we grow flowers

As we entered our thirties we moved into homes with tiny postage-stamp gardens and Marianne became the proud tenant of a 30m/100ft allotment. So we had already been growing flowers for a few years when the opportunity presented itself to rent a glasshouse at Wolves Lane and grow on a larger scale. We jumped in, somewhat foolishly given our inexperience. We agreed to three core sustainable parameters that would form the foundation of our ethos and what Wolves Lane Flower Company stands for: we would grow seasonal flowers without chemicals, we would only work with other British-grown stems when we needed to supplement our own stock, and all of our floristry would be free of floral foam. Initially these guidelines were a response to those niggling floral-industry problems. We couldn't convince every florist to only work with truly seasonal flowers, or at

least buy imported stock from Fairtrade farms; we didn't have the reach or the voice. What we did have was naivety and ignorance about how farming, even micro-farming, works, which allowed us to embrace organic regenerative farming practices without machinery, systems or any idea of just how much compost we would need. Yes, we run a business with an environmental message and practice, but fundamentally we need to sell flowers for that business to work. What we learned through the first year of trial and error, a lot of failures and plenty of tears is that by putting the environment and biodiversity at the forefront, we benefit in more ways than just being able to pay the rent.

We had terrible pest problems for the first few seasons, we still do with certain crops, but working without chemicals has without doubt benefited the ecosystem at Wolves Lane. Our roses and geums were absolutely infested with greenfly throughout the spring season of 2018 and 2019. We wiped the pests off but really it was a losing battle until 2020 when the ladybirds found their way back to our scrap of land which they had evidently crossed off their viable-habitat list. We were looking after nature and over time nature started to look after us. It only takes one tangible, meaningful connection with the natural world to see how important our actions are to the environment around us. It sounds so obvious but nature is so easily forgotten or sidelined in the city.

Growing our own seasonal blooms meant that we had the luxury of working with exquisite, wonderful flowers simply not available at the market or wholesalers. Lots of flowers don't travel well, aren't profitable enough in a voracious capitalist system, or are simply unknown to the farmers in Kenya and beyond. We, and all other small-scale flower farmers around the world, get to choose what we grow and arrange with. We do have to work with unknown climatic conditions but we're never hampered by the availability of a bloom at the market; we just cut what we want to include in a vase or bridal bouquet. And because we grow what we love, we nurture that crop and nurture the place where we grow it – perhaps not as much as our children, but not far off!

We're better florists for not using floral foam. Anyone can stick stems into sponge but every time we need to build an installation, we have to do it in the time frame and conditions of the project in question. In not using foam, we become better at our craft, we limit landfill and protect marine life from the damaging effects of floral foam.

The geum crop, which had previously been covered in aphids.

Growing flowers offered us a lifeline from our relationship with the screen, but more importantly it gave us a connection with the natural world that made us better environmentalists and custodians of the planet. A heightened observation of the natural world is addictive. Turn on your awareness and the more you'll notice the possibilities for discovery are limitless. The climate crisis leaves even the most ecologically conscientious of us paralysed and ostrich-like in the face of depressing and horrifying world events. But focusing on our immediate vicinity, how we interact with it, how we care for it, is something that has helped us to find a way through the sense of helplessness.

The natural world is not something we can visit or dip in and out of, we are part of a complex web of life and have a vital role to play in that web to ensure our survival. Without the insects and the plants we simply don't exist. Flowers are something we are irresistibly drawn to and turn to at the milestone moments of our lives, at births, marriages and deaths, to connect with an estranged friend, to send love or say we're sorry. They colour our most formative experiences and are our gateway to finding our own personal relationship with the planet we inhabit.

Don't worry about the slow progress of your garden or that you haven't yet nailed 'year round colour' – there will always be ebbs and flows in the flower garden. By growing from seed, saving that seed, taking cuttings, dividing perennials, buying bare-root roses – all things we'll cover in this book – and swapping plants with other plant lovers, you can slowly and relatively inexpensively build up your garden collection. It is a life's work but does not mean you need a massive budget.

Embracing seasonality can be tough at first, especially if you're a fan of the blowsiest blooms. Peonies, dahlias and roses are not always in season, regardless of what the flower market would have us believe. However, as soon as you start growing and nurturing your own plants you will notice that any hierarchy of focal flowers is rendered entirely redundant when they start to bloom. Believe it or not, an armful of wild carrot cut from our plot brings us just as much joy as a vase of our garden roses.

SOIL

Autumn is the start of the gardener's year for us. It's also an excellent time to get to grips with some fundamentals of healthy soil. We were both fairly indifferent science scholars at school but have begun to realise how vital an understanding of some basic soil and plant science is to those beautiful blowsy flowers you can then proudly plonk on your bedside table.

Soil is alive. And understanding how it works is crucial if you want to get the best out of your plot. When we were first let loose at Wolves Lane, neither of us knew very much about this crumbly, ancient and ubiquitous matter. And to date a David Attenborough-like champion, who can unlock its secrets and really get people fired up about how vital it is, is yet to step forward.

The first April day we got the keys to the plot, the grasses, brambles and bindweed reached our knees and shoulders. We picked up forks and diligently dug it over. It felt great. We had our very own 'makeover moment' looking at our transformed plot and we vowed to maintain this new sense of order. A couple of weeks later (still with no plants to fill these new beds) the plot was covered in a soft fuzz of freshly germinated annual weeds. We were busy wrestling a new section of the plot and didn't pay much attention. By the end of May, the jungle of weeds was reclaiming it.

This is a common cycle with new gardeners. So why does it happen? While bare soil psychologically feels ordered and controlled, in turning the soil you have brought to the surface a whole heap of weed seeds now only too grateful to you for giving them access to the light to germinate. You have also radically damaged – potentially destroyed – the habitat of the microorganisms that live there. Not being able to 'stay in control' often leads many of us to seek out more 'efficient' methods. You might dig again. Perhaps you start googling tips to more quickly eradicate the weeds that are making your gardening endeavours impossible. What about just one big dose of Roundup? A blow torch to burn the suckers to death?

At this point you need to stop and remember the most vital and neglected fact about soil. Soil is fragile and teeming with life. And the majority of those organisms living within it are the secret to your healthy soil. If faced with an intimidatingly weedy plot, rather than setting to with the garden fork, consider a no-dig approach (see pages 112–115) and cover as much of it as you can to keep the soil protected and the weeds in check while you organise your beds a bit at a time.

THE BASICS OF SOIL

Soil contains a mix of minerals and a complex web of organisms that support its vitality. All organisms require food to thrive and the key – whatever kind of soil you have – is to encourage these hospitable conditions with the inclusion of lots of organic matter into your soil.

What is organic matter?

Organic matter (or humus) is formed as organic materials are converted by microorganisms into a decomposed state. These materials include leaves, plant stalks and the carcasses of the creatures (and their poo) that lived off these plants. All living things are slowly returned to the soil and the soil benefits from the trace

Key Tasks

Attend to your compost heap • Mulch • Sow cover crops

nutrients and structure this matter provides in its broken-down form. Hold a handful of homemade compost and it looks nothing like a commercially created compost that can be uniform and fine in form, but is full of the remnants of twigs, bits of snail shell or half-rotted leaves. The transformation of organic material to organic matter happens at different rates and is a constant work in progress within our gardens.

Within and atop our soil are a whole host of creatures that are hard at work adding organic material to it that will decompose. A grazing cow will excrete and provide a vital food source for a dung beetle. An earthworm collects dead leaves and plant material from the soil's surface and pulls it down into the soil. Large mammals, invertebrates or arthropods (ants, spiders or millipedes, for example) are all easy to spot, but

We invest heavily in our soil for a productive crop.

> 'There is a whole world under our feet begging to be explored.'
>
> Nicole Masters, author of *For the Love of Soil*

there are millions of other creatures vital to this process that are too microscopic for us to see. Staggeringly, there are more microorganisms living in a teaspoon of healthy soil than there are humans living on the planet.

Microorganisms

Four of the key groups of microorganisms that carry out the work of breaking down organic matter are fungi, bacteria, nematodes (non-segmented worms that prey on fungi, bacteria and other nematodes within the soil) and protozoa (single-celled organisms that prey on bacteria). You might not have heard of the last two, and the first two sound like things we spent our childhood trying to avoid or kill – we have been conditioned to think of fungi as poisonous or alien, and bacteria as the reason we get sick. In fact, all four of these organisms play a vital role in making the minerals and nutrients in soil available to plants to allow them to thrive and prosper.

How microorganisms and plants work together

We all know the term 'photosynthesis' from school, the process by which plants access carbon dioxide and sunlight from the environment around them to produce sugars and carbohydrates (i.e. energy) to live. But plants also need more complex nutrients to thrive such as iron, boron, phosphorus, calcium and potassium. These minerals may be present within the particles of the soil (the inorganic elements of it that originally, perhaps thousands of years ago, were to be found in rock form) but the plants need help to access them.

First the tasty carbs and sugars plants acquired through photosynthesis are made accessible to the soil's fungi and bacteria. As more food becomes available to these organisms this creates a microbe baby boom, and this new population sets about harvesting the nutrients present in the soil minerals or 'parent matter'. Further up the food chain the nematodes and protozoa suddenly have a glut of fungi and bacteria to consume and then essentially excrete the nutrients previously consumed by their prey, making them finally available to the plants themselves. When this process works successfully it results in an abundant crop of whatever you're growing – potatoes or pansies – and maximises the fruits of your harvest.

To ensure this process can happen successfully, organic matter is critical within your soil to keep a healthy balance of all these microorganisms. If you are composting and returning organic materials to your soil, this miraculous process will take care of itself.

Air and water

Healthy soil also needs air and water. Both worms and microorganisms are busy tunnelling through the soil to process organic matter and in doing so they create its honeycomb-like structure. More air pockets in the soil means more space for storing droplets of water that plants can utilise when the weather is dry, and more oxygen for beneficial bacteria. With a more porous soil, plant roots can also tunnel easily through soil to create strong networks and don't expend energy working against compacted soil.

Soil is fragile and easily damaged. Heavy machinery, feet or overgrazing by animals can squish this honeycomb network of air pockets, but also rain (which can hit the soil typically at a rate of 15-25mph) can also contribute to compaction, which occurs commonly when soil is left exposed. With heavy rainfall, compacted soil has nowhere for the excess water to be stored. Not only will soil become more easily flooded but it also creates the right conditions for anaerobic bacteria. While not all anaerobic bacteria are bad, on the whole they're not great news, producing acid compounds (often the reason a sludgy heap of compost can smell really bad) and conditions that plants don't favour and can eventually die in.

WHAT KIND OF SOIL DO YOU HAVE?

There is no right or wrong type of soil, but understanding roughly what yours consists of will help you to look after it, and all types can be improved by working in lots of organic matter. Soil structure is determined by the predominant type of soil you start off with: sand, clay, silt or loam.

Sand

Sand particles are large, meaning water can pass quickly and easily through them and the soil is in danger of drying out quickly and can then be more vulnerable to wind erosion. Those busy little microorganisms help with this too, producing a sticky compound that binds these sand particles together to create a complex structure of micro-aggregates that can retain more moisture. Fungi also have some gluey properties and create a network of threads within the soil, binding together aggregates to become macro-aggregates and improving structure.

Clay

In clay soil, the primary particles are referred to as platelets (they literally look like plates under a microscope) and can easily stack up with very little space between them, creating inhospitable conditions for beneficial bacteria as the soil is easily compacted and saturated with rainwater. Lots of organic matter is needed to encourage our friends to get to work and aerate the soil. Clay soil is rich in minerals but plants rely on a thriving community of microorganisms within it to be able to access it as described above.

Silt

Silt soils are made up of particles somewhere between the size of clay and sand. They are fertile, light but moisture-retentive, and easily compacted.

Loam

Loams are mixtures of clay, sand and silt that avoid the extremes of each type.

It's easy to tell what kind of soil you have simply by digging a hole and rolling the soil together in your hand. A clay soil will stick together, sand will refuse your efforts to manipulate it into a ball, silt is similar to clay but a little less extreme and loam will make a satisfactory slightly crumbly ball. Loam is what gardeners want most of all, a combination of all the elements and often the most user-friendly and hospitable to a wide range of plants. All types will benefit from more organic matter, and mulching and composting should be part of your seasonal cycle regardless.

The law of return

If you can, compost. Whatever is taken or grown from the soil should be returned to it. Trees do this effortlessly and brilliantly, shedding their leaves to rot down and create leaf mould – the perfect medium for germinating seeds. Every time we bundle up our green waste and turf it out on the pavement we are depriving our garden of these returning nutrients and thwarting this cycle of life. It's not always our fault. Many of us live in densely populated areas with little or no areas for composting let alone time. But it is totally possible to compost on even a micro scale (see page 79 for more on food composting).

HOW TO BUILD A COMPOST HEAP

We often talk about our propagation house – the smaller glasshouse where we sow all our seeds – as the engine room of the business. But really the true heart of any growing operation has to be the compost heap. For the first couple of seasons we were a bit paralysed by composting. There is a lot of advice on how to do it correctly, ratios, how to monitor temperatures, when to turn, what to put in and in what order. Frankly it can all sound like the unattainable pursuit of an alchemist – and incidentally compost is often referred to as 'black gold'.

The good news is that even for the most slovenly of composters, stashing all your weeds – providing there's some twiggy bits as well as the lush green growth – in a black bag and completely forgetting about them will still result in a form of compost. In the past we regularly weeded, bagged, forgot and returned six months later to find the pesky roots and shoots had broken down into a moist compost. Think of composting as cooking. Anyone can make a basic tomato sauce by chopping tomatoes, frying an onion and adding some water. Once you've eaten that a few times though you're eager to improve. Like everything in the garden, don't worry about being a model composter from the get go but just get started and improve through trial and error.

Top tips for starting your heap

Aim for 1m/3ft squared as a minimum size to enable the heap to heat up and break down quickly. If the heap is too small it won't heat up sufficiently and will decompose slowly. A bought or council compost bin is fine too if you want to keep your compost enclosed.

1. Cover Have a tarpaulin or thick cardboard ready to cover the heap to avoid excess water from rainfall sludgifying your mix and to keep light, which will encourage perennial weeds like bindweed into pursuing a second life, out of the heap.

2. Think efficiently If your plot is large, create several heaps in strategic places to avoid a long journey to the compost heap every time you've weeded or conditioned your flowers.

3. Mix your ingredients You're aiming for a mix of 'greens' (weeds, food waste, deadheaded flowers, grass cuttings etc) which are rich in nitrogen, and 'browns' for carbon (stalks of larger

Make sure to mix or layer your composting ingredients.

plants – but not huge branches – wood chip, cardboard or dead leaves). The greens will heat up your pile quickly (plunge your hand into a big pile of grass clippings and you'll see what we mean) and are full of moisture. The browns are helpful to soak up all this juice and to keep the heap aerated. Without the browns you end up with a wet and sludgy mess which will smell bad as it'll be decomposing in the absence of enough oxygen, a process also known as anaerobic decomposition. Without greens the pile will stay cool and break down much slower.

4. Compost instinctively Some cooks need a recipe to follow to the letter and others are more instinctive. Try to embrace your instinctive inner chef when creating your compost heap and don't overthink it.

5. Drown the thugs We throw most things straight on the compost heap but if you're particularly worried about spreading perennial thugs like bindweed or horsetail consider drowning them in a bucket of water for a few weeks and making compost tea (see Spring Soil, pages 115–116).

6. Don't over-green your heap We try not to get hung up on ratios, but you always need more browns than you think. Some experts recommend a 50/50 ratio of nitrogen and carbon elements, whereas The Land Gardeners' fascinating approach that creates a layered 'cake' of compost works on the basis of 60:1 carbon:nitrogen! In the full throes of the season we are all guilty of merrily throwing our weeds into a heap and hurrying on, and it's clear that we're probably all over-greening our heaps.

7. Turn It's important to turn a compost heap to aerate it, to ensure there is a good spread of microbial life throughout the heap, to cool it down if the centre has really heated up and to ensure all the ingredients are mixed well. We aren't diligent composters so tend to only turn our heaps once after around 4-6 months depending on the season and how active the heap has been. Others do this more regularly and diligently so work out what's practical for you.

WHAT IS MULCH?

In the same way that composting can at times feel like a bit of an exclusive members' club with an impenetrable code, mulching is a term that foxed us for a bit. In the autumn plants die back and slow down but the garden is still working hard behind the scenes to regenerate for a new burst of growth in the new year. Think about all those plants you've tended and loved throughout the year and then imagine the endurance test they're about to undertake over winter. The soil – their home – is battered by eroding and drying wind and flood-inducing rain for several months, so autumn mulching is essentially like tucking a warm duvet around your plants and soil to protect them for the testing time ahead. Unsurprisingly one source of this duvet is to be found in your compost heap. The summer will have churned out an abundance of plant matter that you'll have piled up to compost and spread around the garden. Mulch as thickly as you can afford, a layer at least 8cm/3in deep, and prioritise your favourite perennials or shrubs; do what you can do thoroughly rather than doing everything partially.

If you're not able to make enough compost yourself there are several alternatives you can use for this essential task. Head to Spring Soil, page 112, for more information.

COVER CROPS

Another way of shrouding the garden in a duvet is to think about sowing cover crops, liberally scattering seed directly over the soil in early autumn when there is still enough warmth for them to germinate. Clover, mustard and rye grass are all good examples that will protect the structure of your soil, their roots anchoring the soil in place and reducing wind and excess rain erosion. They'll also prevent unwanted weeds from appearing over winter. We've tried covering our tulip beds with clover this year so they can pop up through the fuzzy layer in spring. Once the tulips have been harvested we'll hoe off the clover, and then simply plant the next crop into the bed.

Above: We grow phacelia, which is a green manure and good for pollinators.

Opposite: Poached egg plant is another brilliant green manure and very attractive to pollinators.

SEED

Seed sowing is one of the most joyful elements of gardening. A brown speck burgeoning into a fully fledged plant in one short season is miraculous, and the rush of success addictive. Once you've mastered a few easy varieties and been able to cut armfuls of sweet peas or calendula, it's easy to be totally seduced by the seed catalogues and order everything like a kid let loose in a sweet shop. Growing from seed is a thrifty way of filling your plot but not if you buy more seeds than you have room for or time to nurture. Be selective and remember, there will be future seasons to try new, alluring varieties.

Throw something down on bare earth in autumn or spring and you're in with a chance of some flowers next spring, but for more accurate results we prefer to sow into modules and trays. If you don't have anywhere suitable to start your seeds inside and you want more tips for direct sowing head to page 118. Regardless of your sowing method, it's important to remember to stay flexible and alert to the conditions of your plot at any given time of year. Each seed will have specific likes and dislikes – how much light or heat it needs to germinate, how deeply you cover it in compost, the medium you're sowing into and how much time, attention and water you remember to give the seeds once sown. No matter how exacting you are, a cold snap or heat wave can also arrive to slow everything down or frazzle your fragile new babies. Or tiny snails can stealthily swarm in the dead of night and decimate your baby seedlings. We can never fully control this process of new life, but there are things we can do systematically to help.

TIPS FOR SOWING SEED

1. Choose your moment

There can often be an itchy enthusiasm to begin and wipe away all of the previous year's mistakes by getting ahead with seed sowing. Despite zeal for the task in hand brought about by the arrival of new seed packets, don't rush this process. The last week of summer or first of autumn tends to be when we start our hardy annuals in earnest. If it's particularly warm we might hold off a little to prevent everything drying out too quickly and making growing conditions rather inhospitable. In a mild autumn we sow right through the middle of the season, then stop until after Valentine's Day when the UK has 10 full hours of sunlight.

2. Choose your medium

Seeds have all they need to germinate wrapped up in that little hard shell but they do need a free-draining, light, fine-textured medium to spark into life. We have experimented with making our own seed compost using roughly half coir, a quarter homemade compost and an eighth of each sand and vermiculite, but it's not precise. Like cooking, you develop an instinct for this and can add ingredients until it feels loose, fluffy and free draining. This season we've been more pinched for time and decided to keep things consistent by buying a specific seed compost. Whatever you do, always buy peat-free compost – gardening is no excuse for the destruction of a thousand-year-old fragile, planet-saving ecosystem.

3. Select your containers and modules

We like to have clear control over how many seedlings we're starting so we use a 40-module tray where each plug is 5cm/2in diameter by

Key Tasks

Sow hardy annuals and perennials • Swap seed with friends

7cm/3in deep. While you will want to upgrade your plants into more nutrient-rich soil, growing them in larger modules means they can be left slightly longer in situ before potting them on. If you're on a budget you can sow into old yoghurt pots (with holes cut in the bottom for drainage), toilet rolls (excellent for long-rooted plants like sweet peas), or buy a cheap wooden gizmo to make pots from newspaper. Seed compost isn't cheap, so you don't want to have to fill too large a container, but equally, tiny modules risk the compost drying out more rapidly. If we're more sceptical about the viability of the seed we're sowing – perhaps it's over a season old and we're not sure how well it'll germinate – we use a wide seed tray to sow more thickly across the surface of the compost. This saves time when we're feeling less confident of good germination and

When making your own seed mix, make sure to sieve the compost.

means we'll prick out whatever does appear a week or so later.

4. Sieve
Your seeds want to wake up into an even, fine compost. If sieving all the compost feels time-consuming, you can break up your compost with your hands and first half-fill the seed trays with this before sieving the top layer more finely.

5. Check moisture levels
Your compost should be consistently moist but not sopping wet – you shouldn't be able to squeeze the water out of a handful!

6. Check light levels
Light is essential for bringing on stocky, sturdy little seedlings and not spindly, sickly specimens. Glasshouses, polytunnels and cold frames are ideal but if you are really pinched for space choose your sunniest window and set up a table where you can rotate sown trays. You'll have to pot them on and move them outside to a sheltered spot quicker than those sown in better light, but hardy annuals will survive temperatures down to a couple of degrees below freezing. Raise your pots onto shelves or tables to make it harder for slugs and snails to find them if possible, and avoid very windy, exposed spots where plants may dry out rapidly.

7. Settle the soil
Fill your modules and smooth your hand over the top of the surface to make sure all compost is distributed evenly – it's easy to leave the holes at the edges less full. Once full, lift the tray up by an inch and drop it two or three times. This allows compost to settle down into any air pockets under the surface, then top up the modules so they really are full and so the soil doesn't sink below the surface of the module when it's watered.

8. Write your labels before you sow
We've found that writing our labels first really helps us stay organised about what's getting sown where and means we don't accidentally sow into the same module twice. We use white plastic labels and write the date and name of the seed and variety, and then stick the label in so the information is clear of the soil level. We reuse the labels by using both sides and then clean them with rubbing alcohol. Wooden lolly sticks or takeaway chopsticks are biodegradable options but can smudge more easily so keep them well away from the moist compost.

9. Count out your seeds

As a rule of thumb we tend to work to the 'heir and a spare' rule of two seeds per module, so you have two chances of success. With smaller, inexpensive seeds like Shirley poppy, you'll inevitably end up sowing more as they're so fiddly to handle.

10. Check soil depth

Small flower seeds are happy to be surface-sown and if they're chunkier we sow at about double the depth of their size. Foxgloves, poppies and nicotiana will want to be just on the top whereas cerinthe, scabious or sweet peas need to be tucked in further. Always follow the instructions on the packet and cover with fine vermiculite to prevent damping off.

11. Water!

Always finish any seed sowing by making sure the tray is thoroughly watered. The safest way is to place modules or trays in a separate tray without drainage holes to soak up water from below, which means you're not disrupting or, worse, washing away the seeds waiting on the surface of the soil. Leave the seeds in water for around 20 minutes and don't forget about them! We often set an alarm on our phone; if you've moved onto another garden job you might go home, leaving them in an airless soggy bed all night.

Once watered, we place all our modules on a reservoir of sand lying on top of staging which is lined with heavy-duty plastic. The sand retains moisture better than the capillary matting we used at first and means the plants' roots nestle into it, easily accessing moisture if we're a bit slow to move them on. You may not have room to set up this sort of system but try to organise something where the water doesn't immediately run onto the floor. Lining a flat surface with a few layers of used compost bags can work.

As we sow hundreds of seedlings, soaking them all in trays after that initial sowing would take too long, so we subsequently water from above with a rose attached to our hose, trying not to blast seedlings too aggressively. If you use a watering can, move this along the row quickly, ensure the holes provide a fine spray only and position the rose so it flows at its slowest rate with the end of the rose turned to face the handle of the can.

12. Have patience

Seeds have erratic germination times based on the conditions when you're sowing them and their individual characteristics. Most will germinate within 21 days but some can be slower so don't give up on them too quickly. If you're nervous and you have time it's always worth doing another batch two weeks later, remembering that the autumn window for good germination of hardy annuals is maximum 6-8 weeks long.

Aftercare of your sown seeds

Water regularly, watching the weather, and keep Goldilocks in mind, who would want her trays neither too wet nor too dry but just right. In very warm spells we water the seed trays every day but usually every two to three days is enough. Keep an eye out for predators – we regularly run our fingers around the edge of the module trays to check there are no tiny snails hiding there ready to strike. We don't struggle with rabbits, mice or deer like our rural colleagues but you'll need a plan to deal with whichever pests you're liable to attract. For our ongoing battle with warmth-loving foxes who break into our glasshouse and bask on the trays (or just poo on them), we are considering covering the window vents with chicken wire to stop them jumping in.

Potting on

Once the first set of 'true leaves' have arrived you'll need to keep a close eye on your seedlings to check that they're not getting overcrowded, leggy, or struggling against any competition in the module or from a lack of nutrients. We pot on once we've got a couple of sets of true leaves but have been known to leave it for longer if we're particularly stretched juggling our floristry work or other gardening jobs. There's more about how to pot on successfully on page 128.

Overwintering

When you sow in autumn you'll need a plan for overwintering your plants. Lots of hardy annuals are pretty robust and can be planted out (see Tend, page 46) and left to weather the elements in all but the most exposed and brutal of growing spaces. They can look a bit sad and bedraggled during the winter months but many will put on new growth and recover well come the spring. Having some backup plants kept in pots in a sheltered greenhouse is ideal if you've the space but most of us don't. In the ground by late autumn is actually often a safer bet than pots left in the garden as plants can really get their roots down before their winter dormancy arrives and will grow stronger than those left in pots. Remember, from late winter onwards you can sow again and plug any gaps.

Seed germination can be erratic, although most will germinate in 21 days.

What to sow now

Some hardy annuals actively seem to prefer an autumn sowing over a spring one so we prioritise ammi, larkspur, orlaya, and – waiting a little later into autumn – sweet peas. We also sow bupleurum, cornflower, snapdragons, malope, calendula, corncockle, gypsophila, chamomile, scabious, stocks, bells of Ireland, cynoglossum and clary sage.

Storing your seed

At the point when you're poring over the seed catalogues and filling your basket, how and where you're going to store all these seeds tends to be far from your mind! Nothing is likely to make you feel more disorganised than a disarrayed box of seed packets in which it is impossible to track how old anything is, what you've sown and when. It's worth spending some time to devise a system that makes sense for you. As we order our seeds from a variety of sources we have now taken to placing all the packets into identical brown office envelopes with the variety written on the front. These are then stored in alphabetical order in four categories: Hardy Annuals, Half-Hardy Annuals, Biennials and Perennials. These envelopes are much easier to keep in order in a veg crate and do help us fight that general sense of chaos that the season can throw you into once there are lots of jobs to do at once. Always write the month and year you received the seeds on the packet too so you can keep track of how old the content is. Seeds need to be kept dry, cool and dark so we also place a waterproof bag around the whole crate to ensure it's not accidentally leaked on, and put it in a back room on a high shelf away from vermin. Sometimes we take some of the seeds that favour a period of cold stratification (larkspur are the main culprits) and place these in the fridge for a couple of weeks before sowing.

During your frenzied seed-sowing period try to keep a list of any varieties you run out of and will need to replace before you begin sowing again come mid- to late winter. Seed suppliers can be quite stingy with certain flowers, whereas you can end up with enough cress, cornflower or rocket seed to last you a decade. Consider a seed swap with other flower-growing friends to make use of excess seed while it's at its freshest and save yourself buying one of everything.

Some crops, such as larkspur and corncockle, prefer to be autumn rather than spring sown.

Store seed carefully and always write the year on the seed packet.

TEND

As summer's blooms start to fade, autumn might seem like a fallow period for flower growers. Although the year's floral productivity is gently winding down, the list of jobs for the following season's harvest is just revving into gear. Once the August bank holiday is over, a certain focus descends over the flower farm to prepare for the next year. Embrace it, and accept it is virtually impossible to ever prepare as much as you aspire to. Whatever hard graft you do put in through autumn will pay dividends in a season full of flowers from spring onwards.

In addition to starting our hardy-annual seed sowing, early autumn is when we start thinking when to chit corms, dig up and divide dahlias, plant bulbs and mulch. There's often a sense of time dissolving away as autumn progresses and days get shorter, colder and wetter.

All through autumn is when we focus on potting on our healthy seedlings, pinching out leading stems before planting out into their final growing positions. See Spring Tend, pages 126–129, for full instructions.

PLANTING OUT

In autumn we plant out hardy annuals inside the glasshouse and in outside beds. The key to successful overwintering is healthy well-established plants which will be best equipped to withstand the winter elements and marauding wildlife. Once we've potted on our seedlings, we allow them a month or so to grow and establish, planting from mid- to late autumn, depending on the weather and how much growth the plants have made. We don't have a hard and fast rule for a final planting deadline. Ideally plants will have a few weeks of temperatures comfortably above freezing to establish before winter sets in. Before planting out, we leave the plants outside in a sheltered spot for around five days to acclimatise to being outside after their cosy glasshouse beginnings. Many gardeners advocate gradually building up their tolerance to the elements by taking them in and out over a few days but we never have time to manage this, and a sheltered spot seems to work fine.

Wait for plants to establish before planting out.

Key Tasks

Pot on • Pinch • Plant out • Sprout and plant corms • Lift or mulch dahlias • Plant autumn bulbs • Divide perennials and gift extras to friends

When planting out, always water your young plants and give them a handful of compost. Remember how traumatic it is to move house? Your plants feel this trauma too, so look after them and give them a large enough hole for the compost and the roots to sit in, then anchor them in place by pressing down firmly around the base of each plant so they can grow straight and strong. Staking won't be necessary until early spring.

CORMS

Ranunculus and anemone corms can be fussy but if you can get them to sprout, you're halfway there. We chitted our first corms in 2018 and it wasn't until 2021 that we achieved a really good crop of ranunculus, but they are absolutely worth the hassle. Most anemones pop open by late March, a time in the year when we've been so starved of colour that we coo over each one with the kind of excitement that we normally reserve for blowsy 'Hot Chocolate' roses or 'Jowey Winnie' dahlias. In spring, rose season still feels very far away and you'll be so grateful for the effort you made in autumn when you start to see the swan-like heads of the anemones unfurl.

Pre-sprouting your corms is not essential but does give additional confidence if you're nervous about wasting space on dud corms.

How to pre-sprout your corms

1. Corms arrive in early autumn. We keep them in a dry dark place until the middle of the season, then place them in a bucket of water and leave them to plump up for 2-4 hours.

2. If you plan to have specific varieties planted together you need to label the buckets and to soak your corms separately – all varieties of octopus-like ranunculus corms or knobbly anemones look the same.

3. Spread the swollen corms out on seed trays on a mix of compost and vermiculite. Anemones are best placed point down (the opposite to a bulb) and ranunculus' little legs should be facing down.

4. Cover the corms with more compost so they are totally submerged, and sprinkle briefly with a watering can using a fine rose. Don't over water or they may rot.

Swollen ranunculus corms after soaking.

Pre-sprouted ranunculus corms.

5. Cover the seed tray with something to block out the light. Check the corms regularly – the sprouting process can take a couple of weeks.

6. When the corms have visible pale shoots and small white hairs they're ready to plant. We always pot up into good-quality compost before planting out as we've found the squirrels and foxes dig the corms up if they're planted too early, but you can plant them directly into their final positions in early winter.

7. We have found that ranunculus and anemones both thrive with some winter protection so we build caterpillar tunnels over them, which seems to encourage good stem length when they start to flower. These are rows of metal semi-circles about 1m/3ft wide and 60cm/2ft high covered with reusable polythene or fleece, but it's not strictly necessary to cover them. The key thing to remember is that they need a spell in cold ground to flower really well – you can wait to start them in spring but if they miss a cold snap they never perform well.

DAHLIAS

Between seeds, corms and planting out you have just about enough time to deal with the dahlias! Tender perennials, dahlias do need to be looked after during the winter months but the cosseting required depends on your location. If, like us, you have a sheltered plot that doesn't flood over winter you can leave your dahlias in the ground for two seasons and dig them up in the third. If they're not lifted and divided every few years you will end up with foliage-heavy plants and fewer blooms.

We decided to leave our dahlias in the ground for the first time in autumn 2019. Marianne had been lobbying to leave them in for some time but the thought of losing multiple dahlia tubers to the wet soggy winter had felt too risky. Eventually we concluded that the amount of time and effort it takes to lift and store them all is just one more exhausting job than we need at one of the busiest times of year. The cost of buying new tubers or rooted cuttings against the weeks that it takes to get all the dahlias lifted and tucked away for the winter made it a risk worth taking. To protect them we mulched with straw, covered the crown with a plant pot and drove a bamboo cane through a hole in the pot to keep it in place and to tell us where exactly we should be looking for new growth in the spring. February 2020 was the wettest on record so we waited very anxiously in the spring for their little green shoots. Luckily we were rewarded with a 90% survival rate and while some varieties had withstood the damp better than others, ultimately we couldn't have asked for a better turnout.

Lifting, dividing and storing dahlias

If you're growing dahlias on heavy clay prone to waterlogging, in an exposed plot or where temperatures throughout the winter regularly reach sub-zero, you'll probably need to lift them. It's not complicated, but it is time-consuming.

1. You must have a failsafe labelling system in place when the dahlias are in full bloom. Once the plants have been cut down and the tubers have been dug up they'll all look the same, so tie labels securely on each one.

2. After the first two frosts of the autumn, cut all the browned foliage right down to the crown leaving just a few centimetres of the thick stalks uncut. It's absolutely fine to lift before frosts but remember that you'll have to keep them frost-free for longer once they're out of the ground and more vulnerable to frost.

Lift dahlia tubers in the winter if you regularly have sub-zero temperatures.

Stored dahlias can be divided and re-potted in the spring to create new dahlia plants. *Dahlia* 'Cornel Brons'.

3. Carefully dig up the tubers using a fork, trying to get underneath the clump and not pierce the tubers.

4. Remove as much soil from the tubers as possible in situ. Then use a hose or outdoor sink to wash all the excess soil off. It's really important that you do this thoroughly so no damp soil is left in any small gaps. If there are any rotten tubers, cut the rot out straight away using a sharp knife or secateurs.

5. Take your tubers inside, somewhere dry and frost free like a garage or shed, and hang them upside down to drain any moisture from the stalks – they can be left like this for about two weeks.

6. Dividing large tubers before storing them can make them easier to move and pack but you can wait until spring if you're short of time. Dividing tubers is an easy way to create more plants but there can be variations between the divided stock and parent plant, whereas cuttings taken in spring are more likely to give you reliable versions (see pages 120–121).

7. When dividing, you are looking for the 'eyes', the little nubs at the top of the tubers that signify new growth. Take a sharp knife and divide your tubers in half or quarters, making sure to discard any that don't have eyes, as these won't provide new plant growth.

8. Store tubers somewhere frost-free, dark and dry. We use vegetable crates lined with cardboard, kraft paper or newspaper, storing the dahlias by variety and covering them in spent compost as a layer of protection. Other growers use sawdust, sand or newspaper to cover them.

BULBS

Autumn is the time to plant narcissi, tulips, fritillaries, alliums, muscari and any other spring bulbs. All bar tulips can be planted from mid-autumn, but wait another month to get your tulips in to ensure the ground is cold enough to avoid tulip fire and other diseases. Plenty of gardeners have admitted to planting tulip bulbs as late as midwinter, but they do need a cold snap over winter.

You'll always need more bulbs than you think. If you're low on budget but want a real show of spring colour, consider planting up pots and troughs in the garden which you can move to the forefront to enjoy them in bloom rather than aiming to fill a flowerbed, which can get expensive very quickly. If you're growing in your garden, plant in blocks of at least 15-20 for maximum impact otherwise they can look a little lacklustre. Daffodils are perfect for naturalising, but tulips will generally degrade in size and quality year on year. At WLFC we treat tulips as annuals and pull the entire stem up when harvesting including the bulb, which gives us taller stems (see Spring Harvest, page 141).

In our first season at WLFC we enthusiastically dug trenches for our tulips, we shovelled London clay for hours and felt incredibly satisfied once we'd managed to tuck all the bulbs up for the winter. Several seasons on and tonnes of organic matter has been added to those beds, making the soil crumbly and easy to work, but trench planting still means a lot of shovelling. Instead we mainly plant tulips into raised beds or crates. Crates mean mobility: you can move them into the shade or just out of direct sunlight if you find all your tulips opening at the same time. With raised beds we tend to find that the soil levels will have sunk dramatically during the season so we can pretty much lay the bulbs straight on the surface of the compost and then cover with new compost. It uses a lot of compost but takes a fraction of the time of the other planting methods.

If you're short of space, tulips can be planted in crates.

Bulbs are perfect for small gardens as you can really pack them in, spaced just slightly further apart than eggs in an egg box. The larger the bulb, the deeper it should be planted. As a general rule of thumb we tend to plant two to three times the depth of the bulb. Spring flowers really are a novice gardener's best friend as all the goodness is in the bulb: you don't need to treat the soil or do anything technical other than plant the bulb at the correct depth, with the point facing upwards, and from early spring onwards you'll be rewarded with the most exquisite introduction to the flower season.

We do suffer from squirrel thievery and fox antics after bulb planting so we cover beds with a layer of chicken wire to prevent them getting dug up. We place metal road pins between each variety and tie the label to the pin so we can keep tabs on which section is which tulip. Tulips flower at different times – the earliest starting in early spring and the latest appearing towards the end of the season if the weather hasn't been hot. We try to group them into order of flowering so we can harvest them most efficiently but you may choose to mix this up if you're hoping for a spread of garden colour.

DIVIDING PERENNIALS

Dividing is one of the easiest and cheapest ways to propagate plants. Over a couple of years, perennial plants will start to grow outwards; they spread and put on new growth on the outer edges of the plant while the middle of the plant ceases to be productive. By dividing up the outer ring of plantlets and discarding the central part of the plant you're encouraging new growth and more prolific flowering for the following season. Autumn and spring, when temperatures aren't too hot or freezing, are the best times of year to dig up and divide perennials that have got too bulky. If you don't have the space for all the new plants you'll create, friends and neighbours will be delighted to take them off your hands!

Key spring to summer crops for us are geums, which repeat flower from spring to midsummer at WLFC. To divide these successfully we use a very sharp garden spade or knife, although some people prefer to use two forks back to back. Dig up the plant, using a fork to lever it out carefully from the ground. Shake off excess soil while leaving some in place to protect the root system, and then use the sharp spade to cut the clump up, discarding any brown bits. You're aiming to

create clumps that are no smaller than your fist. If it's been particularly dry, water the area first so it's easier to dig the plants out of hard ground. Aim to have a space in mind to replant them straight away but you can get away with leaving them somewhere frost-free with a good amount of soil around the roots for longer if needed. Established perennials are tough so don't worry about damaging the root system too much – they'll recuperate quickly. Always water the planting hole generously and mulch with good-quality compost when planting the new clumps. Remember the analogy about moving house – this will also be traumatic for the plants but they will come back happier and healthier.

Dividing perennials is one of the easiest ways to propagate plants.

HARVEST

Autumn sees us continue to cut bucketfuls of flowers including dahlias, sunflowers, rudbeckia, cosmos, zinnias, amaranths and basil. The productivity of the plants begins to wane as autumn draws on and it can be agonising knowing when to call it a day and uproot the plants to make room for next year's crops. By mid-autumn, many plants are looking tired and we know we'll have our work cut out with mulching and bulb planting so we tend to be quite cut-throat and compost plants even when they're still blooming. It all depends on how much space you have and the extent of your ambitions for getting ahead for the next season. Chrysanthemums are a late star of the cutting garden which don't tend to hit their stride for us until the end of autumn. We get the best results by growing them under glass but it is totally possible to select some of the earlier varieties and plant them in the garden too.

Chrysanthemums have an incredibly long vase life, which is probably one of the reasons we're so used to seeing them on the petrol forecourt, but don't be hasty to dismiss them. There are glorious varieties of Spider and Allouise chrysanthemums, among others, in mouth-wateringly deep autumn hues which look fabulous with the last of the rosehips, dahlias and any skeletal dried flowers or seed heads you've preserved over the summer months. Some of our favourite varieties include 'Pip Salmon', 'Avignon', 'Misty Bronze', 'Coral Reef' and 'Fleur de Lis'.

We cut chrysanthemum blooms when they're about three-quarters open. To extend the vase life, cut about 2cm/1in off the bottom of the stem every few days and they will keep well for 2–3 weeks. Chrysanthemums are perennial but can get a little woody and unproductive after a couple of years so we take cuttings every spring to reinvigorate our stock (see pages 120–121).

We do not grow enough perennials on our plot, mainly because we've never felt certain how long we will be able to stay and root at Wolves Lane. Four years on and we kick ourselves throughout the season that we have not invested more in perennials, particularly in autumn when the half-hardies are beginning to fade. For sheer productivity, adding some perennial asters and grasses is an excellent place to start to provide yourself with some invaluable autumn fillers, or you can find yourself very dahlia heavy without enough of a foil to pair the heavy blooms with. We have plenty of self-seeded wild asters at Wolves Lane, which we plunder, but we are currently eyeing varieties including the lovely white *Aster novi-belgii* 'White Ladies'.

Hydrangeas are autumn stalwarts which will flower prolifically for years to come given partial shade and rich, moist soil. We planted *Hydrangea arborescens* 'Annabelle' and *H. paniculata* 'Limelight' two years ago, preferring paniculatas to the more blob-like mop-headed *Hydrangea macropylla* varieties, and although the plants are still small we've been able to harvest plenty of blooms for bridal bouquets. Hydrangeas also provide an excellent bridge between fresh and dried ingredients and they dry brilliantly if cut when mature – when the petals are becoming almost leathery – and left in about 2.5cm/1in of water. You can just hang them upside down to dry if they're really on the turn.

Key Tasks

Harvest dahlias, chrysanthemums and final half-hardies, hydrangeas and late-flowering perennials • Forage for berries • Harvest mature seeds and store • Finish drying flowers and store • Make wreath bases

'[Dried flowers] are the bones of flowers – the skeleton form of them – and their beauty, though entirely different, can in some cases be as great as the beauty of the flowers.'

Constance Spry, *Garden Notebook*, 1940

Watch out for rosehips in the autumn and always leave plenty for the birds.

Annual artemisia or sweet Annie is a lovely aromatic ingredient that works brilliantly fresh or dried. On our small plot we're not always spoilt for choice with foliage and although artemisia takes all year to bulk up, we would not be without it as a beautiful filler for late autumn arrangements. *Verbena bonariensis* is less alluring but still infinitely useful through early autumn, producing masses of flowers for fresh or dried arrangements.

ROSEHIPS AND BERRIES

In autumn the foliage palette slowly transforms from green to russet, gold and red, and rosehips and berries are exciting additions to our arrangements. Some roses are better for hips than others and often those that produce the most beautiful blooms don't develop useful rosehips. The ones we cut the most are from a wild, single variety which was planted many years before we arrived at Wolves Lane. The flowers are beautiful but have limited cutting value so we enjoy them when they bloom in midsummer and then hold out for the profusion of hips in the autumn, leaving plenty for the birds to enjoy.

Pyracantha is rather a municipal, unsexy plant but the long trails of scarlet berries are glorious and can be thinned if they prove too heavy for a vase. Crab apples are a brilliant addition, while sloe, cotoneaster, rowan, hawthorn, privet, snowberry and ivy are all great if you can forage them responsibly. Never eat a berry unless you're confident it's edible – many of these are poisonous.

SAVING SEED

Don't forget seed for next season in your harvesting. We save seed throughout the year, but autumn is the time to save later successions of hardy annuals and half-hardies. Using seed saved from your own plants is incredibly rewarding as you run through the entire cycle of life without needing backup from any seed producers. Moreover, seeds that your own plants have generated should be much happier and more willing to germinate than anything that you can buy, having adapted to your unique growing conditions. F1 hybrids won't produce plants that will generate seeds and are only good for one generation of plants (see page 161).

Keep an eye out to spot when seed pods start to brown and dry so you can collect them before they disperse. Big podded flowers like sweet peas are easy to pop open and harvest whereas it can be frustrating to try to extract the dust-like contents from the seed cases of nicotiana. If seed pods are still green you can dry them on a windowsill until they have turned brown. Some varieties, including cornflowers, need to be cleaned or sieved so that the chaff (the dead bits of the plant) doesn't get saved along with the seed. Some flower heads and pods will need some gentle encouragement like shaking, rubbing or tapping to extract the seeds. Hold the varieties in question upside down over a piece of fabric to catch the seeds. Once you have cleaned the seeds of any unwanted debris, store them in a labelled envelope inside an airtight container where moisture won't damage the viability of the seeds.

THE DRY STORE

This season really is the last hurrah for any stems that will be worth drying as autumn brings increased moisture and humidity, which damages flower heads and tarnishes petals. Snip all the remaining stems of strawflowers, statice, amaranth and grasses before the weather turns damp. If you're an allotment owner, don't forget to keep an eye on your neighbours' compost heaps too! We snaffled a whole heap of wonderful dried corn and perennial gypsophila seed heads from fellow growers who were happy to let us raid their heaps. The sun is less fierce by the start of autumn so we sometimes hang stems in the glasshouse and worry less about a loss of vibrancy in favour of the stems drying more rapidly. If your dry store is a shed or garage that isn't 100% waterproof, make sure everything is boxed up and stored in an airtight container to keep moisture out. Stored in dry conditions, your preserved flowers can last for years like a crispy archive of ingredients to dip into throughout subsequent seasons. More details on setting up the dry store can be found on pages 131 and 167.

The winter damp on our draughty site means that if we leave our dried stems hanging for long they will all start to moulder and the damp will penetrate into cardboard boxes. To combat this we pack into plastic storage boxes, but if you're drying in a centrally heated house or well-insulated room you may be lucky enough to be able to leave stems hanging or you can simply wrap them in paper and place them in cardboard boxes. Keep in mind that proximity to a radiator or bright sunlight will cause stems to fade more rapidly and become brittle. Always label the outside of the boxes – you'll damage the stems if you keep having to rifle through them looking for your nigella seed heads or that prized apricot statice.

Make sure to store all dried stems in a waterproof container to protect against rot.

MAKING WREATH BASES

We have made a lot of wreaths in our time and admittedly in the earlier years most were constructed on wire bases, easily bought from a florist wholesaler. However, once that wreath had come to the end of its aesthetic life, the whole thing had to be dismantled to separate the wire base and reel wire from the foliage, dried ingredients and moss. We now make entirely compostable and biodegradable wreaths on our own bases.

One of the old boys who used to grow on the plot where you'll now find our dahlias was a huge fan of grape vines which he grew to make dolmades (stuffed vine leaves). The vines are wonderful but become totally unruly and invasive if we don't stay on top of them. In the autumn we cut long lengths of the vine stems to make wreath bases. We remove the leaves and make a circular shape starting with the fatter end, weaving the rest of the tendril around itself until the skinniest bit of the vine is tucked in at the end. It doesn't matter if it's not perfectly circular as other lengths of vine can be woven in to even out the shape, and an organic shape anyway tends to suit the material better than something totally symmetrical.

If you have vines but find them too woody to use you can soak lengths in water to make them more bendy and malleable. Alternatively, if you've been too keen and cut the vine lengths a bit green so they snap, don't worry, just keep adding lengths of vine on top and then leave them to dry – they will hold their shape. The key to making a good base is to weave enough lengths of vine to make it sturdy enough to take the weight of all the ingredients, so think about weaving between three to eight lengths of vine of about 1m/3ft long. The vines will usually keep their shape but you can secure the circle by wrapping them with twine at each quarter to help with stability.

If you don't have a grape vine, you can use other climbers such as honeysuckle, clematis, wisteria or Virginia creeper – even bramble if you can face removing the thorns! It's worth experimenting with any branches or materials to see what will succumb to being bent into a wreath shape. We've even had success with branches of forsythia. Willow is an easy option but we prefer the wilder forms of other vines. Store your bases in the same conditions as your dried flowers until you're ready to use them.

A dried-flower wreath with a woven base made out of grape vine.

DAHLIA STAIRCASE

Mechanics
Sphagnum moss
Chicken wire
Reel wire
An assortment of glass vessels
Kenzan (floral frog)

See page 208 for the flower varieties we used

One of the easiest ways to decorate a staircase is to take an assortment of vessels and run them down the steps, one after the other. The flower heads will cover the container on the next step up, so it's a very easy way of creating a vertical floral-scape without getting too technical. Think about the size and shape of the vessels; if you're a beginner, it's always easier to arrange into those with narrower mouths as they support the flowers better.

To give the impression of an organic line of flowers, you'll want to hide your mechanics. Consider the height of the vessel against the height of the steps and place the containers at intervals onto the treads. You won't necessarily need one vessel per step and placing them asymmetrically can create a wilder aesthetic. Work from top to bottom down the staircase to avoid kicking the arrangements over as you go. As soon as you put the blooms into the vessels you'll start to hide the containers above. Don't be too concerned with arranging the flowers – as soon as you start layering the vases in front of each other on each step, you can move the flowers around to create shape and negative space. When you get to the bottom step the last vessel will be exposed, so either use trailing foliage to cover it, or choose ingredients that are happy to be out of water and place them in a flower frog to sit on the bottom step.

In the image on page 66 you'll notice the sense of abundance in the top right-hand corner. We couldn't achieve this using vessels because the steps turned 90 degrees and there was no tread for them to sit on. Instead, we made a mossage, a technique Jay Archer introduced us to many years ago. Essentially you create a sausage made out of chicken wire and stuffed with moss, which provides enough hold and moisture for flowers to sit in without containers of water (see page 67). For very delicate stems you can also insert test tubes, if necessary. When making mossages remember that the flower stems will increase the overall size considerably. To estimate the size, calculate that the installation will be at least two thirds larger than the mossage, so start small. If decorating your staircase with hundreds of dahlias seems a bit much you could copy the technique to decorate a corner of a deep windowsill or mantelpiece with just three or four vessels.

Mossages: sausage-shaped pockets made out of sphagnum moss and chicken wire, which are key to sustainable floristry.

To create a mossage

1. Cut a rectangular piece of chicken wire and fold it in half, long sides together.
2. Attach the short sides to each other by folding cut ends in on each other to create a pocket with three sides closed and one long side open.
3. Stuff the pocket with moss, packed densely enough to hold the stems but not so tightly stuffed that the stems can't get through. Once the pocket is filled, seal the last edge again by folding the wire in on itself.
4. A mossage has no significant weight so it has to be attached to something or weighed down. For this installation we used a long length of florist's wire to attach it tightly to the door at the top of the staircase, but you could tie it to a vertical banister rail.

WI
NT
ER

WINTER

WHEN THE FLOWERS ARE GONE

As the colours slowly fade from the garden from mid-autumn to Christmas, many of us hit a busy spell. We may be focused on mulching, sowing and tidying, or just busy getting on with life, Christmas planning, shopping and dodging the colds doing the rounds. But once Christmas is scraped into the bin and cast out on the pavement, we are really ready for spring to show up and make things a bit easier.

This thirst for colour inconveniently coincides with the flower garden's 'Out Of Office' break. Winter marks a slower, less productive phase, vital for the soil and plants to replenish their energy – just like us – but inevitably means that flowers for picking are in scarce supply. Daffodils can make an appearance as early as midwinter – although most tend to be the primary yellow, story-book type. They smell delicious, but are nearly always more successful left in the garden to flower than plonked in a vase. Crazily, this winter-to-spring smudgy phase includes the two most commercially significant events for florists, Valentine's Day and Mother's Day. Given that our climate and growing conditions provide us with a glut of floral abundance from late spring to early autumn, this timing can feel maddening to us growers.

As seasonal flower farmers we have, to date, not even attempted to provide flowers for Valentine's Day but we always receive a few panic-stricken phone calls in the run up to 14th February, trying to source roses. We try gently to explain that if you're not seeing roses in the front gardens, parks or roundabouts on your commute to work, and you're still cranking the heating up when you get home, that's a pretty strong clue that roses aren't even close to being in season.

So where do the flowers come from that satisfy this flower hysteria in many who don't give plants and petals a second thought the rest of the year? Roses are one of the most lucrative and sought-after commodities in the cut-flower trade and are exported from the space-age glasshouses of innovation and efficiency in Holland, also increasingly from countries close to the equator such as Ecuador, Colombia, Kenya and Ethiopia, which have plentiful and cheap land, labour, water and sun. Even if they begin their lives in Africa or South America, the majority of these stems are then air-freighted via Holland to be bought in the gargantuan auction house Aalsmeer – the fourth largest building in the world covering a staggering 518,000m^2/5,576sq ft – and then transported by road hundreds of miles to the florist who will sell them finally to the consumer days later.

The heavy carbon footprint of a rose flown thousands of miles across the globe seems obvious. But due to the abundance of natural heat and light in Kenya, for example, the carbon footprint of a rose grown there and transported to the UK is less than one grown in Holland – 2.407kg for a Kenyan rose versus 2.437kg for a Dutch-grown stem whose production requires far more energy consumption. A bunch of bananas for context is 0.5kg (see Glossary, page 194, for more on how carbon footprints are calculated). Sea freighting is increasing, particularly from South American countries, where stems are refrigerated at the farm of origin and not opened again until they arrive at the Dutch auction

house, reducing a reliance on air freight, but this is not common practice yet. It is also too simplistic always to assume that British-grown is the answer. British is not synonymous with organic, and flowers may still be grown with artificial heat, light or pumped with chemicals to reduce pest damage. Getting to the facts about where and how our flowers are grown is incredibly challenging, particularly if you consider that the bunch of flowers you pick up in a supermarket rarely bothers to even list a country of origin – a legal requirement on the fresh produce we eat.

Working practices, from environmental and sustainable policies right through to the working conditions for those growing the flowers, are also staggeringly varied. The 2009 film 'A Blooming Business' followed three workers as they live with past and present traumas from working on flower farms in the Naivasha area of Kenya. Examples included facial disfigurement from exposure to the toxic chemicals used to spray the crop, habitual sexual abuse of the women workers at the hands of farm supervisors, extreme job precariousness and even blacklisting if employees complained. This film is well over a decade old but it would be naive to assume all practices like this are now completely in the past. As consumers, we simply don't know where our flowers come from, who grew them and under what conditions. And as we stand at the supermarket shelf, making a spur-of-the-moment, quick purchase, it doesn't occur to us to find out, to ask the questions. Who would we ask anyway?

While growing and cutting from your own garden will circumvent these knotty questions altogether, clearly a billion-dollar industry is not going away overnight. One way of seeking assurance about the growing conditions of your flowers, if roses at Valentine's Day feel non-negotiable, is to opt for Fairtrade flowers, a consumer-facing accreditation that provides some confidence about where and how our flowers were grown. These are to be spotted in UK supermarkets but account for only 2% of imported stems into the UK.

Many pick up a plastic wrap of roses from the supermarket on their commute home on Valentine's Day. It's as easy as buying a packet of chewing gum – a quick convenience – and signifies a microsecond of thought to symbolise your love to a significant other in your life. Instead, why not slow down, plan ahead, order a box of plants from a local nursery to plant when the milder temperatures arrive, give a packet of seeds and the promise of a summer of cut flowers to follow, buy or make a pressed-flower artwork, order a dried-flower bouquet from a florist – easily posted across the country as a non-perishable product – or if you simply must have roses, make sure they're Fairtrade.

Fairtrade

Fairtrade is an accreditation that expects its farms to continuously strive to improve their environmental and social working practices. Workers receive a fair wage and farms receive an extra 10% for every stem, known as the Fairtrade premium. This is invested back into workers' education, community infrastructure, funding to provide vegetable gardens for workers to grow their own food or access to training about workers' rights. Farms can also decide to use part of the premium as cash payments, equally distributed among all workers on a farm. Although pesticide use is rife in the flower and plant industry, the Fairtrade Standards prohibit the use of the most hazardous pesticides. Workers are required to receive proper protective equipment, training in pesticide handling, and regular medical check-ups when working with pesticides. The fact that this is not a basic requirement across all growing operations really gives pause for thought.

SOIL

Winter is a month for noticing bones. The skeletal silhouettes of last summer's bounty can provide a dramatic and beautiful contrast to the greys and browns of the garden. Plus they make statuesque offerings for the kitchen table. And as the tender giants of the summer, the dahlias, zinnias or amaranths are zapped by the first frosts, the verdant abundance draws back to reveal to us the structure of the garden itself.

REDISCOVER YOUR BONES

Pathways throughout your plot and borders provide order, structure and some clarity on exactly how much space you have at your growing disposal. A soggy start to winter can leave your plot rather crumpled, bedraggled and sodden, with you itching to have a good tidy up. This is certainly a fallow time in the gardening calendar so an obvious time to do it, but do keep in mind that some of this 'mess' is vital habitat to wildlife sheltering from the harsh weather and can also provide protection for the soil itself. So don't go crazy. It's for this reason we like to focus now on remaking paths and the grid-like structure of the plot.

We have experimented with black plastic weed membrane for pathways but find for our small plot, it isn't the answer. Weed membrane is of course problematic as a plastic which will eventually end up in landfill, plus it can easily fray and moult into the soil, it's unsightly and it degrades over time, leaving holes and areas exposed where weeds can find the light. For larger flower-farm operations it may feel like a time-saving solution but we have been much happier using cardboard and wood chip to create our pathways.

THE WONDERS OF WOOD

It's always satisfying when you can cultivate a cyclical sense of return within your garden. There is something immensely pleasing about using organic materials that have been thrown away and can now be repurposed and utilised for the benefit of the garden. Our wood chip is all delivered to us for free by our local tree surgeons after trimming back errant North London gardens, but it is vitally important to investigate the source of your wood chip.

Wood chip comes from trees which store carbon. As the wood chip breaks down it can seek nitrogen from the soil to help it process this carbon and help it decompose. Because of this, and because wood chip encourages fungal life which should be kept in balance with other microorganisms in the garden, we tend to stick to using it as a path material rather than a mulch for the plants themselves. If you do want to use wood chip for mulch, let it compost thoroughly for at least a year, or stick to well-established shrubs and perennials, avoiding your shallow-rooted, nitrogen-hungry annual plants.

Creating wood chip paths follows the same principles as other no-dig bed building. Lay cardboard, remove all traces of plastic, wet the cardboard (although we sometimes skip this step) and then add a thick layer of wood chip over it. It is amazing how quickly adding these wood chip stripes through the plot creates a sense of morale-boosting orderliness to a bleak winter plot. You can create a neat-looking path and boundary single-handed in an hour, with the added satisfaction of knowing that over time this material will break down into humus to enrich

Key Tasks

Wood chip paths • Mulch • Leave wild areas for overwintering insects • Install water butts • Utilise food scraps for compost

your plot too. Once paths are marked out you quite quickly get some perspective on how much growing space you'll have to work with, so when there are no sprawling plants to contend with get the tape measure out and do some planning and dreaming about what will go where in the forthcoming season.

More mulching

If you weren't able to get all the plants mulched through the autumn that you had intended to, it's fine to continue mulching into winter until the ground gets really frosted and frozen. After that you should wait until temperatures warm up as the microbes working away in your compost heap, manure pile or bought-in mulch may not survive being disturbed and thrown onto frozen, dormant ground.

Harnessing the elements

As jobs slow down through winter it's an excellent time to make sure you have the infrastructure for your next frenetic growing season as well organised as possible. For us, this includes adding new water butts to the site to capture the winter rainfall. Water is one of the planet's most precious resources and it is becoming increasingly important to preserve and value it as climate change begins to bite. Any roof or surface that rain runs off can be used to capture water – roofs over compost heaps, garden sheds, greenhouses or the side of your house. Always place your butt on a purpose-built stand or a couple of old pallets so you can add a tap and draw water easily from it.

Food waste

While the garden's output of weeds and green waste will slow right down in winter, you'll still be generating weekly waste from your cooking scraps and food. Rather than bagging this up and sending it off in a truck to an anonymous processing plant every week, you can harness its potential to add organic matter to your garden. In 2018 the UK generated 9.5 million tonnes of food waste (equivalent to 25 million tonnes of greenhouse gas emissions), 70% from households. At the heart of the waste issue is, of course, the necessity to buy less and discard less but for all those kitchen scraps, inedible stalks and peel, household composting really needs to become mainstream. It's an incredible resource of useful organic matter and goodness for our gardens.

Several years ago we both invested in a hot composter to live in our gardens and process our food waste. These are not cheap to purchase but provide an ongoing, free resource of organic matter to help temper our heavy clay soil, and mean our homes are closer to a closed-loop system for waste. Hot composters are a bit like compost heaps on steroids: big insulated boxes in which temperatures can rise rapidly if the right mix of greens and browns are added regularly. Theoretically this means you can be using your compost in the garden in as little as 90 days.

It has taken both of us a little time to get to grips with getting the mix right. During a busy working week, it is easy just to throw the contents of the kitchen waste caddy into the hot bin and hastily close it but this is more than likely to make an unappetising, sludgy, stinking form of compost because there isn't enough air and structure throughout the mix to help it break down aerobically. As we have a glut of wood chip on our site, we tend to take home a bucketful regularly to add into the mix, and supplement this with shredded paper, egg boxes, torn up cardboard and brown packaging. Anything biodegradable that will soak up excess liquid and add some air pockets within the mix is brilliant.

We're also just beginning to experiment with the bokashi system, which is a neat and rapid process for turning food into compost, originating in Japan in the 1980s. Bokashi is a process of fermenting waste food by harnessing beneficial anaerobic microorganisms by adding special bran inoculated with good bacteria. You can purchase the magic bran to kick-start the process online, along with two sealable buckets. Within two weeks you can have useable compost, but as with all composting systems, it takes practice to perfect bokashi composting.

Finally there are wormeries, another compact system for processing food scraps. You can buy one as a kit, or make one yourself from scratch.

HOW TO MAKE A WORMERY

1. Use a dark plastic box with a lid, measuring about 50cm/20in wide and deep – any smaller and you'll run out of space quickly for your food scraps.

2. Drill holes in the bottom, sides (at the top only) and a smaller number in the lid of the box to allow oxygen in. You don't want too many holes in the lid if the box will be outside exposed to rain as this may waterlog your box and drown your worms.

3. Line the bottom of the box with some layers of newspaper to prevent your worms falling out.

4. Place your box on some bricks to ensure air can circulate and place a tray underneath to catch any excess liquid. Think carefully about where you place your box as inevitably some liquid will leach out, which may smell or attract flies or vermin. Putting the bricks on grass or putting a layer of wood chip underneath can help soak up excess moisture.

5. Add a first layer, around 5cm/2in deep, of well-aerated organic matter – preferably from another compost heap or wormery – which will allow your worms to settle in.

6. Add some worms! These can be purchased from online suppliers, a fishing shop or preferably get some from an existing heap. The earthworms in your garden are different beasts entirely to those you need, which are brandling worms, *Eisenia fetida*.

7. Begin to add well-chopped fruit and vegetable food scraps. Avoid citrus skin as worms avoid it too. You can use small quantities of cooked food scraps, but uncooked scraps should make up the bulk.

8. Add some browns. Like all compost heaps, a wormery needs an equal amount of carbon to balance the greens and this will help absorb excess moisture from the food scraps.

9. Add your scraps regularly every two to three days. If you overload the wormery the worms won't be able to keep up and the food will begin to stink.

10. Keep an eye on moisture levels: the liquid present in your food scraps should be more than adequate but top up with a quick sprinkle of water if it feels very dry.

11. Keep an eye on the leachate, the liquid that drips out of the bottom of your wormery. Some people use this as a liquid feed for plants but although it may have some nutrient value, it may just be vegetable ooze, indicating excess liquid in your wormery. This could contain pathogens that might harm your plants so it's best to put your leachate back through the system and make sure you have enough brown materials to soak it up.

The advantage of bokashi and hot composting over other systems is that you can add cooked food, even bones if the temperature of the hot composter is high enough to break them down – something you wouldn't want to do with a conventional compost heap for fear of founding a food festival for rats and other furry friends. Whatever system you use, be encouraged by the fact you'll be using your excess waste to feed your soil and keeping it out of landfill.

SEED

SEED YOUR DREAMS

Winter is your season off from sowing. There are the odd exceptions such as sweet peas and cup and saucer vine (*Cobaea scandens*), which will happily get going in midwinter, but really these are the months to seed your dreams and plans.

Goals

Where to start? It may be useful to begin by defining your goals for the forthcoming season. Are you growing to supply flowers for a friend's wedding, do you want to create a riot of colour for a charity open garden day, do you want a consistently productive cutting garden across the year to satisfy yourself, to supply other florists, to offer a pick-your-own patch? Are pollinators your priority or plants that have multiple uses? Do you want edibles, plants that are great for dyeing, for pressing or drying? Brainstorm all the ideas and then try to boil it down to three key, achievable measures of success for the season. For example, aim to grow enough flowers to supply three weddings you've already booked in, to do a monthly pressing of seasonal flowers, to run one open day on your plot where the patch is at its glorious best.

Review the previous season

The most useful approach to evaluating your growing season is to keep notes throughout the year of things that worked and any failures that disheartened you. The best time to plan a future crop is to review how it's gone when it's coming to an end, when the tweaks you'll want to make and mistakes you made are fresh in your head. Perhaps you need to feed your first crop of hardy annuals more consistently, add two layers of net to your chrysanthemums or label your dahlias more diligently. Try to schedule these resolutions loosely into a planner for the year ahead. There is an impossible amount of multi-tasking in gardening and a year's wall chart that you see on a daily basis may help keep you on track.

Take a seed inventory

In early winter, you'll clearly be able to see any gaps from your autumn sowing. Late winter isn't too late to plug these gaps – your plants may flower a little later but most hardy annuals will give you good results from a spring sowing. Make a list of your priority varieties to sow by winter's end as well as plotting out the year beyond this more thoroughly, thinking about half-hardy annuals, herbs, biennials or perennial varieties too. Don't buy any more seeds until you've worked out roughly where everything is going to get planted. It all feels bright and hopeful when you're adding gorgeous photos of blooming flowers to your shopping cart, but when you receive hundreds of seed packets it can get overwhelming very quickly. If you didn't have a chance to organise your seeds properly in the autumn, do this now, making a list of seeds you need to replenish – for example we find the number of snapdragons or rudbeckia in a seed packet is often a bit scant and these will need replenishing quickly.

Order dahlias

Like your seeds, the best time to review your dahlias is when they're in full bloom. You don't have this luxury in winter but you can look back at photos and work out if you have any gaps in your preferred colour palette. We are often a bit light on pales and whites as we're naturally drawn to mouth-watering sunset shades. Philippa Stewart

Key Tasks

Define your goals • Evaluate • Take a seed inventory and order • Order dahlias

Above: If limited on time, grow hard-working perennials such as goat's rue.

of Just Dahlias provides a fabulous photographic resource of unusual varieties to try to source via her website and Instagram. You'll need to get in early to nab the most sought-after varieties, so aim to order by early January to get ahead.

What should you invest in?
We had big plans for an amazing flow chart to help pump out digestible answers for growers in all locations, growing on any scale and level of experience, to illustrate what spending to prioritise on your plot. After several cross-eyed attempts, however, we'd completely tied ourselves in knots so have settled for asking a few key questions to help you figure it out!

How long will you garden on this plot and how much do you have to spend?
1-3 years: Focus on seeds – annuals, biennials and herbs, either bought as plug plants or grown from seed. Perhaps add in some hard-working perennials that you'll easily be able to transplant in future such as geums, sanguisorba, goat's rue (*Galega officinalis*) or sedum. Dahlias and bulbs are well worth adding to the mix, remembering that your tulips will function as annuals.

3-5 years: In addition to all of the above, bare-root roses are a must! You may have missed the boat to order these if you're setting to it in the depths of winter but if you can source them, January is an ideal time to plant (see page 91) and we've listed some of our favourites on page 186.

5+ years: With a longer timescale on your plot it's really worth starting with shrubs, perennials and sources of foliage. We love hebe for flower crowns, buttonholes and as filler for installations that can last well out of water. We'd love to have more physocarpus, viburnum, weigela, smoke trees *Cotinus coggyria* and *C. purpurea*, flowering currants and masses of perennial grasses.

How much space do you have?
If you're growing with acres to spare you'll be able to go crazy on your list including shrubs that need lots of space to establish. They're also great windbreaks. But it really is possible to grow some flowers to cut and enjoy at home just with a window box or balcony. We love cress as a sculptural filler; when it's gone to seed, it can be used green

'If you have a garden, however small, if you have access to field, hedgerow, or common, then you are among the millionaires.'

Constance Spry, *Simple Flowers: A Millionaire For a Few Pence*, 1957

or naturally bleached out and dried. Other cut-and-come-again crops like calendula, cosmos or annual scabious will also do fine in small containers. Cram pots and crates with bulbs for a big hit of spring colour and you can even include dahlias if you have substantial pots and can put them somewhere they'll get sun for half the day.

Climbers are your best friend for making the most of vertical space; jasmine is one of our favourite foliages for bridal bouquet 'wangy bits', sweet peas give you a real glut of colour and scent to instantly lift your mood when waking to a bud vase on your bedside table, *Clematis montana* will cope with some shade on a balcony, and *Nasturtium* 'Milkmaid' gives you an edible addition to your summer salads to boot.

Remember that with all container gardening you'll need to water regularly and deeply (with saucers underneath so that the water doesn't just run straight through). You'll also need to feed regularly after the first six weeks from planting, and realise that climbers can look raggedy quite quickly. Deadhead, replenish with new seasonal plantings and move pots in and out of the spotlight to get the best out of your display, moving pots of spent bulbs into a hidden corner to die back.

How much time do you *really* have?

Think realistically about how much time you'll have to sustain and manage your plot. How far away from your house is it? Stepping out of your back door will mean you can keep an eye on seedlings several times a day but may mean you're working with a smaller space than an allotment up the road. If you're time poor, perennials are lower maintenance and more drought tolerant if planted at the right time of year, mulched well and watered deeply as they establish. Consider making the most of easy-to-grow annuals such as phacelia or nigella or use fuss-free perennials like *Alchemilla mollis* or catmint to keep weeds at bay between larger crops such as roses or fruit trees.

Sanguisorba is another perennial that will establish over a couple of seasons.

Grow climbers such as sweet peas to make the most of vertical space.

TEND

The prospect of hosing down tools and greenhouses may feel deeply unappealing when it's cold and damp outside. But this work is crucial. Throughout the year millions of microscopic organisms, pests such as aphids and mites, and bacteria will build up in your growing spaces, particularly under cover. The last thing you want to do is give them the opportunity to survive through winter and ruin all your gardening bounty come spring and summer. For our 40m/130ft glasshouse this is a mammoth task and we try to approach it like removing a sticking plaster – quickly and efficiently!

We use white vinegar diluted with water plus a few drops of tea tree oil in a rechargeable, electric spray bottle to douse all glass, wood and the metal framework within the glasshouse. Fellow growers We Grow Colour gave us a tip to include some horsetail (*Equisetum arvensis*) in the mix, soaked overnight, as it's a weed with great anti-fungal properties. Our commitment to being chemical-free is the backbone of our business and this has just felt like the safest option for us rather than using more abrasive chemical solutions. Once sprayed inside and out we scrub and then hose down with water. We also remove any moss built up in the crevices of the metal frame which will over time cause degradation and leaks.

If you use a lot of containers, empty last year's spent plants and clean and prepare them for next year with the same solution. Don't wait until you actually want to plant up your pot as by spring you'll have a huge list of things to do and are more likely to be tempted to skimp on the cleaning! Terracotta pots risk cracking in extreme cold temperatures so gather unused ones up to store somewhere sheltered over winter.

The same process can and should be repeated for your tools. It's easy to look at a forlorn hand fork rendered rusty and unsightly by being accidentally abandoned in a compost heap and want to discard it and start afresh. As we have got more serious about gardening it's become increasingly clear that buying quality tools that we'll love and want to look after really helps. But if the freneticism of the growing season has left tools caked in soil (our clay soil is notorious for this) or rusty, then leave hand tools to soak in a bucket in some trusty white vinegar or coat in baking soda and a little water to form a paste first. A final scrub with wire wool can help, and don't forget to clean your floristry snips and secateurs thoroughly too!

Infrastructure

Winter is the ideal time to install water butts, create compost systems or clean out the shed ready for the season ahead. Think about how you want your garden to function throughout the year, and what you might need to get organised and have at hand for the busy periods. Have you got enough plant supports, do you need to bulk order seaweed feed or more butts for making your own compost teas, for example? Assign space for these elements where you can get to them most easily.

Plant

It would be a mistake to think that there is nothing to plant in the winter months. A bare-root plant is essentially what it sounds like, a root ball – free of soil – with a couple of stalks at the top which are

Key Tasks

Clean greenhouses, polytunnels and tools • Think about infrastructure • Plant bare-root plants • Prune roses

in a state of dormancy. This is by far the cheapest way to buy your 'investment plants': roses, ornamental or fruiting trees and shrubs, bushes and even some perennial plants which will be useful for foliage or berried stems in subsequent years. It may feel less appealing than buying a lush and leafy tree in a pot from a garden centre but, in fact, the plant is more likely to establish happily when planted in winter. In this dormant phase the plant will be relatively unbothered by any weather conditions and will slowly wake up in response to the climatic conditions that evolve rather than undergoing any transplanting stress. We order by the end of the summer for delivery before Christmas and try to get any bare roots planted by midwinter, or when the ground isn't too frozen or sodden to plant into. If winter turns into a dry spring, then do remember to give your bare roots a good soaking once a week.

HOW TO PLANT A BARE-ROOT PLANT

1. Assess conditions from late autumn to early spring and plant bare roots when the soil is workable, not heavily frosted, snow covered or flooded.

2. Soak the bare root well in a bucket for at least a couple of hours, or overnight.

3. Dig a generous hole two to three times the size of the roots with plenty of space for them to nestle comfortably in. The crown should sit flush with the surface of the soil – too deep and it may rot, too shallow and roots may be exposed and dry out and the plant may rock about.

4. Square holes are preferable to round holes to prevent the roots circling inwards.

5. Remove any stones, perennial weed roots or dead wood, so the soil is crumbly and free draining.

6. Water the hole with about half a can, even if the ground is very moist, before beginning to position your root and back fill. Give a little extra help – rose grower pros David Austin advocate a sprinkle of mycorrhizal powder directly onto the roots of a bare-root rose before planting. Roses may thank you for a couple of generous handfuls of manure or homemade compost in the planting hole, but when planting a bare-root tree Monty Don advocates leaving good-quality compost to mulch around it after planting rather than in the hole. This encourages the plant to spread its roots and search for nutrients.

7. Firm the soil gently around the root to disperse any air pockets but avoid stamping and compacting the soil. Water the planted root again to welcome it to its new home.

8. Mulch thickly with your compost to help lock moisture in, but not right up to the stem.

9. Remember that the first year of life is critical to a bare-root plant. Keep tabs on it during dry weather and give it a really good evening soaking once or twice a week during dry spells, at least one watering can at a time for a bare-root rose and more for a young tree.

10. Stake a tree that will need support straight away while the soil is soft.

11. Weed regularly to remove competition, particularly from strangling varieties like bindweed.

12. Mulch again in late autumn in subsequent years to protect against weather and provide a new boost.

Plant bare-root roses between November and March.

PRUNE ROSES

Pruning strikes fear into the hearts of many newbie gardeners. It can feel like a brutal activity, and as it takes place at a time of year when there'll be no new growth for some time, it can be anxiety-inducing to know if you've done it right or whether you've killed the plant.

There are two bits of good news. Roses LOVE a pruning and they are tough as nails. By cutting your rose back you prevent it from getting overly woody, thick stemmed and leggy. When spring arrives, new growth will appear from your pruned stems more vigorously than if a rose is left unpruned. It is really hard to get it so wrong that the plant will pack up and die, so be confident and just keep a few key principles in mind.

We wait until January to prune our roses. There are fewer jobs to do then so it's easiest to put some proper time aside and do it all in one go. We read one tip that a good seasonal indicator for pruning is to wait until forsythia is flowering. Obviously this is not necessary but is perhaps a cheerful and bright note to self! If for whatever reason you don't get round to it and March rolls in, we'd still recommend pruning then rather than leaving your roses for a whole year.

Wear gloves. Roses are exceptionally thorny and not only is getting pricked uncomfortable but the thorns vary in size and the smaller ones can easily get trapped under your skin. We've had some nasty infections from rose thorns so approach the task armed with heavy-duty gloves or leather gauntlets.

Reduce an established rose by at least a third in height. Any rose that has flowered for two seasons is defined as established. Measure the overall height of the rose and then cut a few stems at a third to half of the height of the rose. Use these cuts as markers. It's not necessary to worry too much about cutting back to a new growth point or cutting on a slant but simply to reduce the height evenly.

Once you have cut everything back, you should then look out for the four Ds, dead, dying, damaged or diseased stems, and remove these too. The experts at David Austin remove all remaining foliage from the stems, because old leaves can harbour diseases from the previous season, though it's not something we've bothered with so far. There are also differences of opinion regarding what to do with your rose prunings after the job. Some recommend burning or sending them out with your green waste collection whereas Charles Dowding asserts that everything is fair game for the compost heap as any diseases will die off within the heat of the pile. If you do decide to compost prunings, cut them right down. We once added 30cm/12in length stems to our hot bin and jammed up the bottom with the stalks, making it extremely hard to excavate the compost!

Prune roses to promote vigorous growth. Reduce the shrub by at least two thirds.

A small note on weeds

We discuss the importance of ground cover on pages 36 and 152, but do keep an eye on your indestructible companion, couch grass, which can creep stealthily over paths even through the depths of winter. While you don't want to leave great swathes of soil exposed, you definitely don't want the couch grass to gain too much of a foothold before the season gets going, so some light weeding is never a bad thing.

HARVEST

If this book teaches you anything, hopefully it will be to embrace seasonality, to appreciate and savour the changing of all the seasons, winter included. The London climate is relatively mild and undramatic so we have little to complain about but the winter can still be a long and bleak season to get through. We make it better by getting out onto the plot on any dry days when some watery sunshine may make it through the cloud. These days make us feel like we've won the lottery and we gladly get on with winter tasks before the light vanishes at 3:30pm. It's the anticipation of spring and that pause before the season starts again that makes winter bearable and enjoyable. It's the interlude between the cycles that allows us to stop and reflect, rest and then make plans for the season ahead. We have friends and colleagues who live in places where their proximity to the equator means that there's very little or no perceptible change in season and we don't know how they do it. We need winter as much as summer – we might not always feel like that but if we didn't have the quieter fallow periods we wouldn't enjoy or appreciate the abundance of summer as much as we do.

And so to harvesting. It's true that while the cutting garden has been put to bed for winter, leaving much in its dormant state, there is plenty of beauty to be enjoyed and gathered at this time. Abundant and varied foliage to create table runners, garlands and wreaths is all there waiting for you in your garden, hedgerows and allotments: you've just got to get out there. And yes, flowers are pretty scarce in the winter but some winter-flowering plants and shrubs are really worth the wait.

While this book isn't about motherhood, we both happen to have had very tiny babies at different times during the winter months. It is a pretty tough season for new mothers and newborns, especially during lockdown, but walking on the dry days and nosying around people's front gardens, observing what was in bud and what was flowering, provided a connection with something reassuring and promising regardless of whatever was going on in our personal lives or the wider world. The seasons will continue to change. There is a security in that which we'll always be grateful for.

WINTER FLOWERS

Searching for flowers in winter has for us often meant we've turned our attention to what shows up in the front gardens, municipal spaces or hedges around where we work and live. *Viburnum tinus* is one of the most common of the lot and while the flowers don't have a huge wow factor, we look at them with different eyes in winter, eyes hungry for blooms of any form. Viburnum is very hardy and long lasting. Use a sharp pair of secateurs to cut the stems – they will happily last over a week in a vase, even with the heating switched on.

Winter-flowering clematis 'Winter Beauty' is a favourite of ours because of the delicate little bells and its clambering habit. It's not one that we cut so much as enjoy while it's in bloom but you will get a day or so of enjoyment from it if you bring it inside. Witch hazels are another rare treat. We don't have room to grow them for cutting on our plot but love the fiery sight of them across London in the early new year. If you are blessed with one in your garden, do treat yourself

Key Tasks

Forage for winter greenery • Pot up and enjoy paperwhites • Make winter wreaths

to a stem or two for some fleeting scented beauty in a vase.

Mahonia, perennial wallflowers, winter jasmine, forsythia, flowering currants, winter honeysuckle, winter box (*Sarcococca humilis*), cyclamen, hellebores and snowdrops are all useful ingredients to intersperse with your winter foliage. Snowdrops or cyclamen may be diminutive and only suitable for a bud vase but can be all you need to lift your spirits on a dull day. As many winter-flowering plants are shrubs, you'll need to invest time and budget to add them to your garden, but winter shrubs can be some of the most precious when you have little else to choose from. We would prioritise flowering currants, winter box (for the scent alone) or winter-flowering honeysuckle if we were starting over at Wolves Lane.

Don't forget the dried stash

By the time the soggy days of winter have arrived your dried materials should all have been safely stashed. Winter is your prime time to get the boxes out and begin to have fun with your crispy treasures. A biodegradable dried wreath (see pages 60–61) is easily made by laying little posies of stems around the edge of a vine base and attaching them with one continuous piece of twine (tied on to the base at the beginning). But dried flowers don't need to be used solely for wreaths. Try them for a mantelpiece arrangement, stick a few choice twisty stems into a kenzan or create something as large and dramatic as our dried-flower nest (see pages 102–103). Revel in the fact that dried stems need no water source so mechanics need not worry you. When you use fully biodegradable mechanics such as straw or twine you can carefully take creations apart and repurpose materials again and again.

Forcing paperwhite narcissi

Some effort is required to force bulbs, but it will pay dividends for the scent alone. As with all bulbs, all the goodness is in the bulb itself so this is a pretty fun and easy project to do for any flower-growing newbie or for a winter project with kids. Allow four to six weeks from planting to flowering for paperwhites. For a Christmas display, that means adding this to your list for November. The same technique works with amaryllis, hyacinths or even grape hyacinths but the pale simplicity of paperwhites makes them our favourites for forcing.

Bleached materials

If you buy in additional dried stock for winter projects be aware of the risks and environmental impacts of using bleached or dyed materials. If you stumble across whiter-than-white stems of ruscus, asparagus fern or grasses in bright jewel colours these will undoubtedly have been chemically treated. It is notoriously difficult to find out much about the provenance of dyed and bleached flowers but they all use chemical processes, some potentially carcinogenic, others coating stems in microplastics to add strength following harsh bleaching treatment, and none is good for the environment. Some people argue the use of these chemicals is acceptable to render the flowers 'everlasting' but once a plant is dyed or bleached it has transitioned from compostable to only suitable for landfill and it ceases to be useful to us. Brilliant results can be achieved using just sunlight to bleach materials through the summer. Our naturally bleached bamboo stems were our favourite new dried ingredient of 2021.

Plant paperwhite bulbs into a mix of grit and compost with bulbs about 2.5-5cm/1-2in apart and the top of the bulb just below the soil surface.

Keep the soil damp but not waterlogged and store the bowl or pot somewhere cool and dark like a shed or garage until pale shoots begin to appear. Once the shoots have grown to about 20cm/8in bring them inside into a sunny spot. Once the paperwhites have flowered, treat the bulbs like any outdoor-grown narcissi that you want to flower again the following year: let the leaves shrivel to allow the goodness to return to the bulb, store them somewhere cool and dry and pot them up again the following year.

SIMPLE WINTER WREATH

Mechanics
Vine wreath base
Straw or sustainably sourced sphagnum moss
Garden twine
Mixed foliage and/or dried flowers
Reel wire (optional)
Length of velvet ribbon

Using the vine wreath base that you made in autumn (see page 60), take wet straw or sustainably sourced moist sphagnum moss and layer clumps onto your wreath base

Secure the clumps using twine, pulling it tight as you add your moss or straw. We recommend twine for a foliage wreath that you may eventually want to add to the compost heap, but for an everlasting wreath that you can keep for longer, wire can make it easier to secure the stems more tightly.

As you place the clumps of moss or straw, wrap the twine around the wreath base multiple times to create a web or series of triangular sections.

Tie off the twine once the wreath form is covered in your base layer.

Create small bunches of a mixture of evergreen winter foliage – take care of anything with spikes and remove any that might give you a nasty surprise when handling later. To make a bunch, layer pieces of foliage on top of each other in a fan shape, with the largest pieces at the bottom. Try a mix of colours and textures. You can add dried elements within largely fresh bunches but remember that these will be more fragile so the more delicate elements might be best tucked in nearer the end.

Once you're happy with your bunch, tie the stems off tightly with garden twine. When you have six or seven bunches start attaching them to your base.

To start, take a roll of twine (or reel wire for dried), cut off a piece about 2m/6ft long and tie it securely to the base. Take your first bunch and insert the stems into the moss or straw and tie this tightly to the base using your pre-attached wire or twine. Don't cut the twine once you have pulled the knot closed.

Take your next bunch and layer it over the previous one with the tips of the leaves covering the stems of the first bunch. Push the stems into the moss and then tie the stems to the base tightly,

as with the last bunch. Continue to do this until your wreath is complete. You may want to use even-sized bunches with similar ingredients or choose to create something more asymmetric with longer stems grouped in one part of the wreath. The final bunch will be smaller to hide the stems of the first bunch.

Attach this last bunch by weaving the twine around a lower layer of foliage and fixing it to the base, then take very sharp secateurs and cut off the stems up to the tying point so it's hidden under the adjacent foliage.

Once complete, review the wreath to decide which position it will hang at. Sometimes turning the wreath round can make the shape work better. When you've decided which are the top and bottom you can add in extra textures of seed heads, berries or dried flowers to finish off if you wish. Dried and delicate ingredients can just be slotted in between the bunches.

Finish the wreath with a long velvet ribbon in a sumptuous winter shade for a pop of colour. We like to leave ours trailing rather than tying a bow.

DRIED-FLOWER NEST

Mechanics
Straw bale
Chicken wire
Wire or twine

See page 208 for the flower varieties we used

You can create much smaller dried-flower nests than ours for your own home and if they're light in weight you can use adhesive hooks to attach them to the ceiling, leaving them to hang cloud-like from above. One of the joys of arranging with dried-flower stems is that they are everlasting as long as they're protected from moisture. So after we had created this nest, the stems were all removed, left out to dry and then boxed up to be used again another time.

You will need something to hang your nest from such as a grid, or beams or hooks in the ceiling or walls. Make sure to check the maximum weight load as you don't want your nest to come crashing down.

To build the nest first create a mossage (see page 67) filled with straw instead of moss as you don't need a moist base. Use twine or wire to suspend the straw pillow safely from your hanging points.

Start by using the larger dried elements to cover the mechanics, pressing them firmly into the straw. Then create pockets of colour and texture by introducing finer elements such as dried larkspur, statice, strawflower or feverfew. It can take a lot of stems to cover the mechanics so for maximum visual impact and ease of creation it's often easier to add bunches of stems together rather than single stems one at a time.

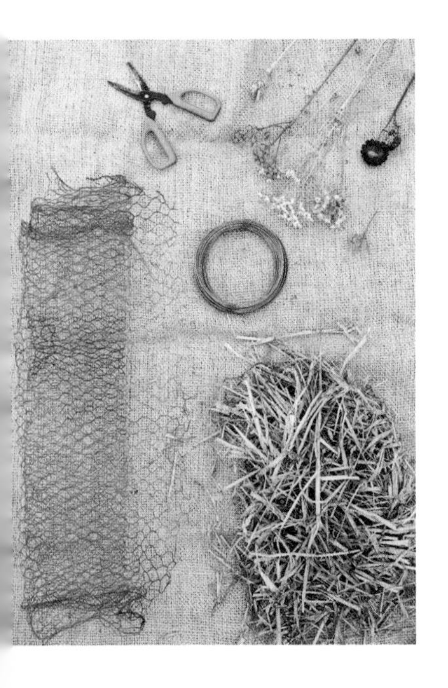

SP RI NG

SPRING

If Valentine's Day sits firmly in the depths of winter, then at least Mother's Day begins to usher in the new buds of spring. We'll confess that early spring is not one of our favourite points of the flowering year. The juicy stemmed blowsiness of the flowers on offer – tulips, grape hyacinth, daffodils – can be awkward vase fellows alongside the corresponding twiggery of flowering trees such as cherry, plum or blackthorn. You can be in danger of creating a rounded little bunch of colour with some wild angles sticking out of your bouquet like TV aerials.

But despite these struggles with spring floral design, these first flowers bring total joy. For a country-born girl like Marianne, Mother's Day was a yearly father/daughter fixture. Up at the crack of dawn (probably 7am in reality!) with wellies over pyjamas, the duo would creep out of the back door and into the garden. Any signs of colour or glossy foliage were snipped with the kitchen scissors and crammed into a vase to present proudly alongside toast in bed to her mum. While neither of our London gardens offers quite the bounty that Marianne's winding childhood garden of apple trees, herbs and veg patch did, we'll be repeating this yearly flower date with our sons. There is always something to cut. Even the most municipal of plants – hebe, winter box, photinia – that grace the fringes of many retail-centre car parks can look fresh and alive if cut and brought into the house on a gloomy day.

As over 55% of the world's population now live in urban centres – and 12% of UK residents have no access to a garden or outside space at all – formative experiences like a welly-clad hunt for spring flowers become less and less commonplace. For a floral pick-me-up to enjoy in our homes, most of us turn to the supermarkets or letterbox flower subscriptions that rely heavily on imported flowers to keep prices as low as possible.

Clearly we haven't always been a global economy, pinging wraps of gerberas around the globe via Holland. As late as the 1970s, UK-grown flowers accounted for up to 80% of flowers bought by British consumers. Prior to this, our iconic florists such as Constance Spry, or early florist shops such as Pulbrook and Gould, would supplement their stock with bounty from the rambling gardens of wealthy estate owners, who would bring them carloads of foliage, climbing roses or mock orange for high-society parties.

So what changed? As early as the 1950s the Dutch government began subsidising energy resources for Dutch farmers, enabling them to scale up their growing operations, work cooperatively to improve efficiencies and invest in research and technology. Over the next decades Dutch growers experienced huge success growing collaboratively at scale. The 1980s saw the arrival of the 'Flying Dutchman' to the UK – Dutch traders who would drive over lorries filled with refrigerated flowers, bringing them direct to florists across the UK and selling at the lowest prices available. Their enormous lorries – dwarfing the tiny florist shops they're delivering to – are still a familiar sight across the UK.

The 1990s saw the rise of supermarkets, behemoth corporations able to secure rock-bottom prices for flowers by buying at scale. With the arrival of alluringly cut-price, straight

With spring comes the anticipation of flowers such as the veined beauty of rocket petals and forget-me-nots.

and uniform stems from Holland, supermarkets saw how flowers could be a cheap marketing ploy to entice shoppers into their stores. Have you noticed how flowers tend to be near the entrance of a supermarket? Or wondered how supermarkets and the supply chain behind that £4 bunch of tulips can make any profit to sustain their business? They don't. Supermarkets use these cut-price, plastic-wrapped bunches as the sacrificial lambs to entice shoppers in to spend money on all the other offers and products in the store.

In some ways supermarkets could be praised for making flowers accessible. A bunch of narcissi can cost £1, cheaper than a packet of sweets, bringing some cheer to kitchen tables across the nation. But this price tag hides the true cost of flowers. To provide flowers fast and out of season requires a particular kind of efficiency – shed loads of insecticides are necessary to guarantee large volumes under hectares of glass, or masses of artificial heat to accelerate the blooming process.

Spring, for us, has always been about anticipation, mentally urging on the arrival of the anemones, tulips or branches of blossom we're so hungry for after winter. There's a slowness to this cycle that intensifies the reward of colour when it finally arrives. We hope that you too managed to plant some bulbs in autumn and will, like us, be continually glancing outside to check their progress. Be aware that as soon as the sun arrives they can all appear at once!

With the arrival of spring comes a new mode of activity. There are seeds to sow, more mulching to do, and perhaps you invested in perennials that can be planted out before temperatures get too high. Drawing on the energy reserved from a quiet winter of planning and dreaming, now is the time to really get going again!

'To some people it is still a matter for comment and surprise to find what really dramatic effects are to be achieved by the good use of the simplest flowers.'

Hostess, Constance Spry with Rosemary Hulme, 1961

SOIL

WHEN TO BEGIN?

By the end of winter we can't fight our instinct to try to get ahead of the season by putting our recently hatched plans into action. The new year begins like a clean sheet of paper and the urge to be your most organised feels irresistible. But if you only remember one thing about growing it should be that every single year will be unique. This will impact when you start establishing new beds, turning the compost heap or sowing seeds. Rather than following a spreadsheet of dates religiously, stay alert to what's really going on, to the temperature fluctuations and particular circumstances of the season at hand. In 2021 we had a cold and dry period in early spring when it's traditionally damp and starting to warm up. It slowed progress on sowing and planting out and left us with a tight window to get things in the ground before summer suddenly heated up. However much you want to get on when spring arrives, make sure you're working in line with the current year's version of the season.

Two of our most important soil 'to do's' in spring are to eke out any remaining growing space by setting up new no-dig beds, and to excavate our various compost piles. Our priority is to mulch any plants that missed out in autumn. One of the key benefits to a spring mulch is to lock into the soil all that winter rain before the dry weather arrives. We like to think of our autumn mulch as soil protection (from rain and wind) and the spring mulch as moisture retention. We wait until the temperatures have warmed a bit before this compost turning and mulching: we're no soil scientists but it feels likely that a good percentage of the beneficial microorganisms nestling at the bottom of the piles might perish if suddenly exposed to frozen conditions, and mulching snowy or frozen ground will be counterproductive and slow your plants' progress.

When daytime temperatures are safely above freezing, spring is an excellent time to establish new beds. We've followed Charles Dowding's no-dig method since attending a workshop with him, and it's easy to start. You'll need a fair amount of organic matter but your soil will be healthier, lose less carbon than rotavated or dug soil and form a richer microbial structure than in traditional 'double dig' gardening. Over time, your no-dig bed will need less organic matter than a plot that's been dug over.

Sourcing organic matter

We're not able to make as much compost as we'd like yet so we use a mixture of an organic mulch made commercially, well-rotted horse manure (at least a year old or it can be too nitrogen-rich for the plants), mushroom compost, leaf mould, spent hops, wood chip or seaweed. Be aware that whatever material you use may come with some risks – there is no cheap magic-bullet compost solution. Always ask about the provenance of what you're buying and what might have been used in its production. If you buy horse manure, find out if they receive organic feed. Is the hay they're fed treated with pesticides? Could there be traces of antibiotics or glyphosate in the muck or compost you've ordered? Foraged seaweed can be coated in salt and will need to be left out in the rain for a while.

It can be hard to gain total assurance about provenance so do remember that whatever you bring onto your plot, once there, may be hard

Key Tasks

Turn compost heaps • Mulch • Establish new no-dig beds • Make compost teas • Weed

to get rid of and may affect the health of your plants in the long term. Also remember to leave a couple of inches gap around the base of shrubs and trees when mulching with manure so that its high nitrogen content won't burn the leaves.

Returning to your heap

A productive and satisfying job for early spring is turning the compost heap. There is no greater evidence of the vitality of the garden and the soil during those seemingly frozen and dead months of winter than when you see the results of the compost heap as spring arrives. Even if the top of the pile looks like it still has a way to go, if it's been sitting there since autumn there will probably be some good crumbly stuff at the bottom. Our heaps are set up as three bays next to each other. We turn the top of the oldest heap into the adjacent bay so those materials now take their turn on the bottom. Check that the heap feels properly layered as you turn it. If it is too wet you'll need to add more carbon-rich materials, too dry and you'll want to water layers as you go. (See page 34 for more tips on building your heap.) Don't start too early on a total overhaul of the heap, however, as you may disturb animals like toads or hedgehogs hibernating in its warmth.

CREATING NO-DIG BEDS

1. Mark out your new bed area. For a dedicated cut-flower bed we recommend 1m/3ft width so all flowers are accessible, but some taller flower-farming friends with more space to play with go for 1.2m/4ft width. Think about efficiency of harvesting – although it's tempting to make a bed as long as possible, once your flowers are a jungly mass of foliage you'll have to walk all the way round every time you forget something or need to take your flowers inside. 4–5m/13–16ft long works well for us.

2. Cut down weedy stems, but don't attempt to dig out all the weeds from the bed area before covering with cardboard. The point of no-dig is that the combination of cardboard and compost will suppress the life-giving light to the weeds and kill them off. By letting nature take its course you're preserving that precious soil structure.

3. There are some heavy-duty perennial weed exceptions to this rule. Alkanet, dock, bramble or horsetail need to be dug up by hand before laying cardboard as they're too mighty and will break through. We value our hori hori knife – a Japanese tool that crosses a knife and a trowel – for these kinds of stubborn weeds, but a strong hand trowel or fork (cheap ones bend and break easily in our heavy clay soil) will do just fine.

4. Lay down cardboard, having first removed any plastic tape that would will just degrade and dissolve into the soil in splintery fragments. Overlap the cardboard so there are no gaps – wherever you leave a gap you'll see the weeds popping up in a regimented straight line where the light has got through.

5. Wet the cardboard to begin the breakdown process before putting your compost on top of it. We seem to be having increasingly dry springs but if you garden in a wetter area you may skip this step.

6. Add an even layer of organic matter on top of the cardboard. 10cm/4in is the minimum depth needed, 15cm/6in is ideal but it is challenging to source enough. Then compress the compost with a board of ply or similar to make sure of your depth. A seemingly thick layer may compress to barely 5cm/2in which would not be adequate.

7. Plant! You can plant straight into your new bed but you need to be a little mindful of the planting medium you're starting with. If using bought-in municipal compost for example, perhaps add a good handful of richer potting compost to the planting holes. The hori hori knife can be handy to cut into the cardboard a little if you need to get the roots firmly planted into the soil, but never make too large a hole or you'll just provide a fringey ring of weeds around your precious transplants by letting light in around them.

Compost teas and additives to the soil

We first trialled compost teas after reading about them in Georgie Newbery's *The Flower Farmer's Year*. Two of the easiest garden plants to use for compost teas are nettles and comfrey. Don't eradicate a nettle patch in your garden; it's your garden larder of useful nutrients for teas, and nettles act as an accelerator to speed up composting when added to your heap (see page 152). Nettles are cherished by butterflies, moths, ladybirds and other garden creatures so they will also promote biodiversity in your garden. The most often cultivated comfrey is the Bocking 14 variety. It doesn't spread so furiously as some varieties but is still packed with goodness.

Jennie Love of Love 'n Fresh flowers and the NoTillFlowers podcast is researching weed juices as an extension of Korean Natural Farming,

which creates beneficial soil inputs from natural ingredients growing on your plot. The idea is that many of our common garden weeds with long tap roots are mining useful nutrients from deep within the soil. Horsetail roots typically go as deep as 2m/6½ft and bindweed roots can be found a staggering 5m/16½ft below ground. If the roots are fermented in rainwater the theory is that they will make good plant food. We're experimenting with buckets of each of our perennial foes plus a more generalised mix of annual weeds to see what works best on our patch.

You can make a compost tea in a bucket or a whole water butt. Steep your weeds in rainwater to avoid the microbe-zapping chlorine in mains water. Cover with a lid and leave the weeds to ferment for a couple of months right up to a year, then use the extremely smelly elixir as a foliar feed. There are different schools of thought about how much to dilute the concoction. We tend to use a ratio of 1:100 for most ingredients but we're still experimenting. Some gardeners think that the actual nutritional value of these homemade concoctions is miminal but at the very least it's a safe way of breaking down some persistent weeds, particularly horsetail. If its silica-rich properties help combat mildew and promote general plant health, so much the better. It's particularly important that horsetail has broken down completely before you use a horsetail tea.

Opposite: We grow comfrey to make a compost tea.

Below: No-dig gardening is all about locking carbon into the soil by adding organic matter instead of digging over beds to get rid of weeds.

It is such a prehistoric survivor we live in fear we'll inadvertently spread it further round the plot!

The power of weeding

We like weeding. It is a mindless and calming activity and provides that hand-to-brain connection that is the cheapest form of therapy available! Once temperatures begin to climb and spring showers arrive, the weeds really begin to thrive. The key is to keep on top of them without spending your entire gardening day weeding. Remember that soil likes to be covered to protect it from erosion and to prevent it from just throwing up yet more weed seeds. So when you clear a patch of weeds do replant the area with something you want. Weed directly into a container as any scrap of a real thug like horsetail or bindweed dropped back into the soil will regenerate and regrow. If you don't have other plans for a weeded patch, sow ground covers such as poached egg plant *Limnanthes douglasii*, red or crimson clover, sainfoin, vetch or even chamomile. We're experimenting with these between bigger perennials, shrubs and roses.

SEED

After a winter poring over seed catalogues the urge to get sowing is strong. Many growers invest in heat mats and lights to be able to jump start the season by a few precious weeks but we wait until Valentine's Day has passed to start our spring sowing. It's an easy date to remember, and in the UK will give your seedlings 10 hours of daylight to get life off to the very best start. Depending on the weather your seedlings may germinate before this but the conditions make it trickier to guarantee that they will thrive, given low light levels and night temperatures regularly plunging below zero. Easy exceptions to this 'rule of patience' are sweet peas and cup and saucer vine (*Cobaea scandens*), which will germinate readily throughout the winter, and some perennials like a period of cold stratification and germinate very slowly.

From late winter onwards we sow our classic hardy annuals in the glasshouse – *Ammi majus* and *Ammi visnaga*, larkspur, wild carrot, *Orlaya grandiflora*, calendula, cornflowers, gypsophila, snapdragons, corncockle and *Malope trifida*, plus any fancy new things that have caught our eye in the seed catalogues. Next in line for sowing will be flowers good for drying, then the half-hardy annuals.

Seed packets can be confusing and sometimes list plants as hardy / half hardy / perennial simultaneously leaving you scratching your head about the right time to sow. Seeds are relatively cheap to buy or free to save or swap so try not to get overwhelmed and just have a go. We experiment all the time. In general, if you start heat-loving annuals too early they will germinate but they'll grow really slowly and ultimately it doesn't really get you any further ahead than waiting three weeks to start them when temperatures have warmed up.

Half-hardy annuals for us, in the warm south of England, can be sown under glass all through spring, sometimes with a direct sowing verging into early summer if the season has been slow to warm up. Sunflowers, coriander and calendula are quick from germination to maturity – calendula can germinate and flower within 60 days – so are worth a late sowing if you need to bolster your plant numbers.

DIRECT SOWING

We sow everything into modules at WLFC as we have a finite amount of space and like to be in control of the number of plants we have per variety. But if you have extra space in a sunny spot then direct sowing is a brilliant way to inject extra colour into your plot and create more potential cutting plants. Direct sowing is best either in early autumn for hardy varieties like cornflower, nigella or ammi, or mid-spring when temperatures are warm enough for easy outdoor germination. At the front of Wolves Lane is an unloved verge of sloping, poor soil which we are slowly transforming into an annual flower patch. We started by covering it with mypex for a few months before removing the worst weeds and gently aerating the area and since then it has proven relatively self-sufficient in self-seeding and regrowing. Watch levels of rainfall when you first sow, as a baking-hot spell will not encourage germination, and sow into uncompacted, weed-free soil then water well. We have found that direct sowing works well for ammi, godetia, feverfew, late-spring sown cosmos, borage,

Key Tasks

Sow annuals • Move volunteers • Take cuttings • Pot up dahlias

campion and cornflowers but you'll discover what thrives in your conditions, which may differ from ours.

If you plan to direct sow into beds you'll need to hoe off annual weeds and take out stubborn perennial weeds. You can prepare a simple 'drill' by running the end of your rake down the bed to make an indentation to drop your seeds into, or measure and place canes at opposite ends of the bed and run a line of twine between them to make sure your drill is straight. Water the shallow channel with the fine rose of a watering can before sowing to provide a moist environment to welcome seeds into. Sow two or three times as many seeds as you think you'll need plants, but it's a waste of seed to sow too thickly as you'll just have to do masses of thinning when the seeds all germinate.

Volunteers
We absolutely love this term used to describe any plants that pop up in spring having seeded themselves from the previous season. As the plants have grown successfully on your plot before, chances are these seedlings will be strong, vigorous and well suited to your unique growing conditions. You'll quickly learn to identify the first true leaves of seedlings to know what to save and what to weed out. Nigella, calendula and ammi are all stalwart volunteers for us but you'll probably find others that germinate easily in your own growing conditions.

CUTTINGS

When you're a newbie to gardening the tendency is to go to a garden centre and buy a basketful of abundant-looking pots – spending a small fortune in the process. It's only once you begin to get some success from sowing seeds yourself that the whole process of growing frugally can take hold. The next natural progression from sowing from seed is to take cuttings from existing plants to expand your stock, getting plants for free! It's another miraculous process that can open up the wonders of plants. Spring is a brilliant time to experiment with your first cuttings, starting with varieties that have high success rates: chrysanthemums, scented pelargoniums, sedum, salvia and willow all root very readily and will give you confidence before you try others.

Easy steps for taking cuttings

1. Take your cuttings in the early morning or on a cool day when stems are turgid. As soon as you take a cutting it's a race against time to get the stem prepared and potted up before too much moisture transpires.

2. Use a sharp knife or scissors to make a clean, straight cut on the stem and cut just below a node. This is a little nobble on the stem that will be the place that new roots will spring from.

3. Rapidly transfer your cuttings into the medium you'll use to root them. You're trying to cultivate new roots before the stem rots or dries out so you want an extremely free-draining mix. Coir, perlite, vermiculite, grit and sand are all good additives to your soil mixture, which should feel extremely free draining and gritty.

4. Cut away any large leaves or shoots that are attempting to flower – you want your stem to fully concentrate on rerooting rather than trying to support a new flower or a large leaf that will transpire moisture rapidly.

5. Bury the bottom node of the stem just under the surface of the growing medium and spray the cutting liberally with water. You can fit several cuttings into one pot by spacing them around the perimeter of a plant pot. Placing them at the edge also helps prevent water loss.

6. Place a plastic bag – a sandwich bag works well – over the cuttings and leave somewhere bright but not in harsh, direct sunlight. Make sure it's somewhere you'll remember to check on them regularly – they'll need respraying occasionally to prevent the cuttings from drying out but not so often that the growing medium becomes sodden, which might encourage the stem to rot.

7. Rooting can take a couple of weeks, evident in signs of new growth or when the stem resists coming away when it's gently tugged.

Our favourite half-hardy varieties

1. Zinnia *Zinnia elegans* 'Queen Lime'
2. Zinnia *Zinnia haageana* 'Jazzy Mixture'
3. Velvet trumpet flower *Salpiglossis sinuata* 'Black Velvet'
4. Amaranth *Amaranthus cruentes* 'Hot Biscuits'
5. Nicotiana *Nicotiana* x *hybrida* 'Tinkerbell'
6. Statice *Limonium sinuatum* 'Apricot Beauty'
7. China aster *Callistephus chinensis* 'King Size Apricot'
8. Winged everlasting *Ammobium alatum*
9. Celosia true wild form *Celosia argentea spicata*
10. Cosmos *Cosmos bipinnatus* 'Xanthos'/'Polydor'
11. Phlox *Phlox drummondii* 'Crème Brûlée'.

For best annual phlox germination cover the seed tray with newspaper to restrict the light until you start to see shoots appear.

Check new tubers for ‘eyes’ that signify new growth.

Pot up firm tubers in peat-free compost.

Make sure that tubers are fully submerged in soil.

Once potted up, label and water once until you see new growth.

PROPAGATING AND PREPARING DAHLIAS

Dahlias are the workhorses of the summer season. Repeat flowerers, strong stemmed and appearing in colours and shapes to meet all tastes, they're one of our absolute favourite flowers of the cutting garden. The work to arrive at this riot of floriferousness begins in early spring.

Every winter we choose a few new varieties to add to the cutting garden, buying new tubers from trusted suppliers which means we have a total of around 180 plants in the cutting garden (see pages 200–201 for our favourite suppliers). We're incredibly lucky to have a smaller glasshouse we rather grandly call the propagation house. It stays moist and warm, providing a temperate environment which allows us to pot up our tubers in mid-spring. If you live in an area where temperatures remain low right into spring, only consider potting up dahlias once the thermometer hits the mid-teens. We start this process when it's about 5°C outside but in the glasshouse it's a balmy 15°C. If we're nervous of plunging night-time temperatures we cover the pots at the end of the day with long sheets of horticultural fleece. Surplus packaging bubble wrap would also do the job. If you're limited for space be realistic about how many tubers you can start this way. A sunny porch is an option but if you're really short of space wait until temperatures are frost-free and only pot up then.

We pot tubers up into good-quality compost (peat-free of course) in pots that are large enough to house the multiple stems that will spring up as temperatures climb. The cost of potting up lots of tubers can be high so it's fine to mix in some spent compost for bulk.

Once you've unwrapped your tubers, check the clump for 'eyes', the little nodules at the top of each tuber that signify new growth. For a newbie, these can be hard to spot, if in doubt try potting them and see what happens. The tuber must feel firm, it mustn't be shrivelled and dehydrated. If there are any sections of tuber that have fallen off the main clump and don't have any little nobbles, discard them as they won't establish into a plant. But any with eyes can be separated and potted up. The potato-like tubers of a dahlia look pretty uninspiring in early spring, but they will give you such a thrill when they establish and bring the most amazing burst of summer colour to your garden.

If you've lifted dahlias from the garden, most clumps will have the previous season's stem visible. We make sure that tubers are fully submerged in the soil and then cut away this old stem, which would get in the way of new growth. Leave your potted tubers somewhere sunny and protected: a conservatory, a south-facing windowsill or south-facing garden wall will do. And water sparingly. One of the major mistakes that we made in the first few seasons of dahlia growing was to lovingly and overzealously water them, which sadly led to a lot of tubers rotting away. Experienced dahlia growers, Philippa Stewart and Erin Benzakein of Floret Farm (see page 201) give their pots a quick sprinkle once potted but don't water them again until they see the first shoots appearing from the tuber.

The dahlia shoots progress into established plants pretty quickly once a thick leading stem emerges. You can pinch this stem (see page 129) – just like you would a hardy annual – right down to the first few sets of leaves. This will encourage more plant growth, branching stems and a vigorous, prolific dahlia.

Dahlias hail from Mexico so wait to plant out until any threat of frost has passed – usually in late spring or early summer – and do harden them off for a week before planting. If you don't have a garden, dahlias thrive in pots but remember to transfer them to a larger pot as they grow, and keep on top of watering and feeding them – they're hungry plants!

At WLFC we try to make sure that all dahlias are planted out by the end of spring, meaning that the new season's stock will start to flower in midsummer. Beware of pests: young, lush dahlia leaves are slug caviar and without protection slugs will decimate your crop. Since we're chemical free we use nematodes, microscopic worms which arrive in a packet in the post as an off-white powder. We just add it to water to make a solution and treat the plants and soil with the nematodes. Other methods like beer traps, copper tape and prickly or gravelly barriers can work but the nematodes are most successful for us.

TEND

CHECK IN WITH OVERWINTERED HARDY ANNUALS

Hardy annuals planted out in autumn can look incredibly sad over winter. A bit eaten, windswept and with no new growth to speak of, it's important to keep visiting them and assessing their progress as spring rolls on. Not all of the plants will necessarily make it through the winter, particularly if you planted them small and late and they didn't have a chance to bed in before winter arrived. If there are gaps, fill them with any surplus plants you didn't plant out in the autumn, which will catch up with their neighbours – you're sure to have sown too many. It's good to check for your own morale as well as for the plants' sake – cornflowers, for example, look like the most depressing addition to a cutting garden over winter, but they stage a remarkable comeback when the weather warms up.

Once overwintered plants are growing away, it's a good time to pinch them out before they get too leggy (see page 129).

Staking

March is the month to think about staking, which is to say, think about it before you really need to. You're trying to beat the sudden surge of growth that annuals in particular put on from mid- to late spring, or you'll have to deal with top-heavy plants nosediving onto surrounding paths and neighbours. It's tricky to prop them up with twine and canes when they've already grown big. We use pea and bean netting which, although plastic, is something we've managed to reuse for many seasons, repairing with twine when we accidentally snip through a square. Jute netting is a good biodegradable alternative. In each corner of our bed and at 1m/3ft intervals (closer in the glasshouse where plants get really very heavy and often scrape the ceiling) we push a metal road pin through the netting to keep it taut. Recommended by fellow flower farmer Paul Stickland of Black Shed Flower Farm, road pins are strong and very easy to push into the soil, though perhaps not the most aesthetic addition to the garden. If you don't want to use netting, sticks of hazel or bamboo also work with strong twine wrapped around them and zigzagged through the bed. This infrastructure will be

Key Tasks

Check overwintered plants • Start staking • Pot on • Pinch out • Weed • Set up a drying area

invisible once the plants shoot up. If you get your staking sorted early – when the ground is still moist and has some give – you can easily push the supports in firm and deep. They will need to hold the weight of lush, heavy flowers come summer and need to be robust.

POTTING ON

If, like us, you seize the new season and sow tray upon tray of seeds, it won't be long before you have a lot of baby plants straining for more space. This can be very season specific. In our 2021 season the seedlings were slow to grow – responding to the cool spring – and then all of a sudden started bolting and growing like the clappers. This is one key reason why we start our seeds off in larger modules or a more substantial seed tray so they can get a little bigger before needing to move home as we are sure to be pressed for time. Wait until the seedling has two sets of true leaves on the plant before potting on. It is also vital to remember that the seed mix your seedlings germinated into was purposefully low in fertility and you'll not want them hanging around when they'll be casting about hungrily for nutrients to put on strong growth. If you've sown an 'heir and a spare' into your modules, it's very likely you'll end up with two per module. The ruthless will just pot on the stronger, but we're not that cut-throat when it comes to seed babies and tend to pot both on for surplus!

Top tips for potting on

1. Water your modules before you begin to move your seedlings to prevent the soil from collapsing. This keeps the fragile root system as intact as possible.

2. Always hold a leaf and not the stem. The leaf is dispensable. Think of the stem as the spine of the plant. Without it, it's game over.

3. Respect the roots. Plant roots are incredibly delicate and throw out microscopic hairs that you want to disturb as little as possible. Discarding your 'spare' probably makes sense so you don't have to painstakingly detangle two sets of roots but if, like us, you're going to do it anyway, take it slowly and gently.

4. Don't be tempted to rehome the seedling in a huge pot. You want to pot on into a comfortable size up for the plant but not so big that the compost around the existing root ball stands soggy with moisture as this can cause the plant's roots to rot. 9cm/4in diameter is our go-to size

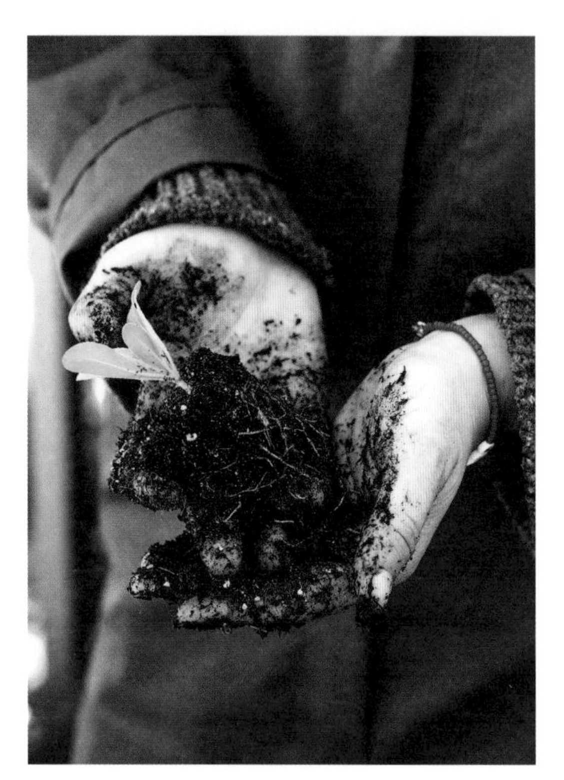

Larger modules allow seedlings to become established before we move them.

Do not be tempted to press down hard when potting on; the root system is fragile.

When pricking out hold the leaf, not the stem.

We pinch out autumn- and spring-sown crops like these nicotiana plants.

pot and we buy the grey or taupe versions which can be recycled commercially if they break.

5. Always, *always*, use peat-free compost. Flowers are wonderful, whimsical and life affirming. They are, however, not worth the total degradation of an ancient and vital carbon-sinking habitat that took thousands of years to form. There are loads of brilliant, high-quality alternatives out there and we've listed a few under Resources (see pages 200–201).

6. Gently fill the spaces around the plug with compost. If the stem was on the verge of getting a bit lanky we usually bury it a little under the compost to anchor it. Don't be too firm – keep those microscopic roots in mind as you don't want to crush and damage them. The plant's roots will find their way through the compost more readily if the compost isn't compacted.

7. Once rehomed, place your pots in a tray of water for 10-20 minutes to have a good drink. Set a timer so you don't forget about them. If the weather is warm and bright, you might want to place your seedlings somewhere a little shaded while they recover from their major transplant. We all find moving house stressful.

8. If you miss the moment with a tray of seeds for potting on – and this happens to us all the time – a tray that's 'gone to seed' and shot up and started flowering can produce lovely additions to buttonholes or for drying flowers in miniature.

Pinching

Pinching is terrifying. You've miraculously nurtured this plant from brown speck to lush, green little plantlet and then, just as it's really getting going you're expected to pinch away all its progress! Why do we do this? By cutting away (we tend to cut with snips versus a thumb pinch to minimise the chances of crushing the precious stem) the central stem, you're encouraging new growth to begin again from this pinch point. You'll be persuading the plant not to channel all its energy into one stem and a larger flower and instead you'll end up with a bushy, branching plant with a mass of flowers. Annual flowers are intent on growing, flowering and setting seed to ensure their succession, so by pinching you're also delaying this inevitable process of setting seed.

The rules of what to pinch and when can feel intimidating. On the whole, *most* annuals respond well to pinching but there are some exceptions. Stocks or non-branching sunflowers for example, flower on a single stem so pinching them out would only result in a decapitated and flowerless plant.

Planting out
This is ruled by the unique seasonal variations of the year in hand and you'll always need to adjust your schedule accordingly. We aim to get spring-sown hardy annuals in the ground by mid-spring but in reality that isn't always practical. If it's been cold and rainy, planting out on scale is an unpleasant task and you're likely to compact the soil structure by treading and sliding all over it.

Before you plant your seedlings you can leave them outside somewhere sheltered for a week to harden off. Many gardening buffs tell you to bring them in and out each day to protect them from plunging night-time temperatures but we don't bother. Our plants are placed outside somewhere sheltered from the wind, raised off the ground to make them less attractive to slugs and snails, and somewhere we'll pass by them daily so we don't neglect them.

Planting out following a decent downpour is ideal. The ground will be easy to work and there will be moisture available to welcome the plants and help them settle in. Do always add a good glug of water to the planting hole, even if conditions have been wet, to ensure the soil further down is moist too. This next big move for your plant will be another stressful shock. Seasonal variations will require you to adjust the attention you administer to your plants. Lots of rain will usher in armies of slugs and snails, delighted to welcome spring with a verdant feast of your nurtured plant babies. Too dry and your plants will find it hard to get going with such small root systems and will need more attentive watering. If the weather is exceptionally windy this can also dry plants out very quickly.

If planting into a no-dig bed, your hori hori knife will come in handy for puncturing a hole in the cardboard. You want to firm in your plants a little more assertively than at the potting on stage as the plants are now bigger and tougher and your aim is to anchor them firmly to help them withstand blowing around in the wind.

Weeding
We discussed weeds a little in Soil but they loom so large in a gardener's life it would be remiss not to include some further notes within this section.

There will be a particular week in spring when the weeds will suddenly begin to perk up and another point later in the season when they will really begin to take hold. The key is to try to spot the first milestone and respond quickly. While some annual weeds surrounding established, overwintered hardy plants are arguably creating some ground cover and are unlikely to overwhelm your flowers, bindweed, cleavers or couch grass are a very different story.

If there is a section of your garden or plot you're not ready to plant into – perhaps you've earmarked it for some half-hardy annual later in the season – then *do* cover it. Cardboard or MyPex either pegged down or weighed down with wood or bricks will prevent the weeds from overwhelming this area before you're ready to plant into it. It may not look beautiful but you'll be endlessly grateful that you did so when, come early summer, you're not confronted with a new jungle to clear before planting out seedlings bursting out of their pots and desperate to get in the ground. When you're ready to plant, remove and fold up the MyPex for future use, or compost the carboard, rake off any dead plant matter and remove by hand any stubborn roots or bigger weeds like dock.

Invest in a good hoe. A long-handled oscillating hoe will help you to check weeds before they get too big for their boots.

Watering

Spring can be a frantic time of multi-tasking. If temperatures heat up early you may feel anxious to get the hose out to bolster your plants. Generally any perennial plants that have been established for a year or more will be very self-sufficient. Overwintered annuals will also cope admirably for a week or so without any additional watering. The plants to prioritise are newly planted annuals: if they are yet to fully establish, a sudden dry spell will check their progress. One good soak once a week (twice if it's unseasonably warm) is infinitely more beneficial than a daily sprinkling. Plants need to be encouraged to spread their roots deep to search out water. A daily shower will simply tell them to root shallowly and send their roots out horizontally near the surface. Avoid sprinklers, which throw water onto leaves rather than getting down to the roots. If you want to ensure plants are getting a good soaking a watering can is probably the most reliable tool for ensuring the job is done properly, but it depends on the size of your plot. We invested in irrigation in our second and third seasons and this has been a game changer. For more about irrigation see Summer Tend, page 166.

If you can, use rainwater. It is softer and doesn't contain the chlorine, salts, minerals and trace chemicals found in tap water, and plants respond better to rainwater. The exception is germinating seedlings; rainwater may contain pathogens the baby plants won't yet be tough enough to withstand.

Looking ahead

Growing flowers demands that you're always thinking three to six months ahead about the tasks you'll need to be ready for. We only work with seasonal British flowers: no imported chrysanthemums, roses or exotics over the winter months means that we have to make sure we're set up and ready to dry as many flowers as possible over the flowering season. From June onwards, jobs in the garden can seem never-ending so early spring is a perfect time to have your drying area prepared before you need it. We are also constantly experimenting with drying spring flowers – ranunculus, muscari and wallflowers are all beautiful when dried.

THINGS TO CONSIDER WHEN SETTING UP A DRYING AREA

1. Optimum conditions are dark, cool and airy with no hint of damp.

2. The drying rack, pole or branch should be easily accessible for attaching hanging stems.

3. Work out where you can store stems once they have fully dried to prevent them succumbing to mildew.

4. Prepare stems at the point of harvest if you can – they become fiddly, fragile and frustrating when it's cold and damp and you have numb, winter fingers! Pack in uniform numbers so you don't have to keep unpacking and checking what you have.

5. Label your boxes well to prevent having to handle fragile dried material unnecessarily.

HARVEST

Late winter and early spring flowers are like precious jewels. Through the mud and the mire a lone, bright face of *Geum* 'Totally Tangerine', a rogue scabious or an early anemone can sparkle back at you, providing reassurance that there will, once again, be flowers.

Early spring is all about being thrifty and resourceful with the flowers you can find, and finding ways to make them shine in all their diminutive glory. A lovely example of an early season jewel of a flower is the rocket flower – fine veined, delicate, long lasting and with characterful wiggly stems. Our flower-starved eyes can rarely imagine a more beautiful flower in early spring. They also last brilliantly in the vase and provide movement in an arrangement when flowers can be quite rigid. *Fritillaria uva-vulpis* bulbs are inexpensive investments that provide some useful delicacy with their nodding heads in an arrangement. They naturalise easily, returning every year with zero effort from you if you don't disturb them through the year.

We are tight on space at Wolves Lane and don't have many wild areas or hedges or verges where we can tuck in bulbs to leave undisturbed, but there is no doubt that narcissi, fritillaries and alliums are total workhorses, turning up to greet you year on year. Invest in more unusual varieties of narcissi that you won't be able to find in the supermarket aisles. 'Delnashaugh', 'Pheasant Eye' and 'Replete' are favourite varieties but there are hundreds to choose from and virtually all smell delicious! Narcissi are best cut when buds are fat and colour is showing but they're yet to open. Alliums should have assumed their full ball form but some of the miniature flowers that make up this globe may be yet to open. *Allium siculum/ Nectaroscordum siculum* is our absolute favourite variety, with dangling bells and excellent stem length. If you leave any alliums too late to harvest you can dry them beautifully.

Ranunculus and anemones grow under cover at Wolves Lane. We grow some under glass to guarantee early blooms – sometimes as early as midwinter – but this can mean they fall foul to aphids and other sap-sucking creatures. Low, outdoor caterpillar tunnels work well for us to get optimum stem length and protect the plants from the worst of the weather but these flowers are cool crops and will cope fine uncovered as long as they're not in too windswept a spot. To harvest ranunculus at the optimum moment the bud should have cracked and softened but not fully unfurled, sometimes called the 'marshmallow stage', when they're soft and squishy but not yet blowsy. Ranunculus last for a couple of weeks in a vase if harvested at the right time. They're also mind-blowingly beautiful as a dried flower if you cut them when they're fully open to preserve their full blowsiness.

Anemones will open up happily if cut in bud and have a similarly excellent vase life. These are one of the key culprits (tulips and ranunculus do it too) for continuing to lengthen in the vase – sometimes even growing a little overnight – so bear this in mind when placing them in a bouquet so they don't end up too proud of the arrangement.

Along with the early spring bulbs, blossom and flowering shrubs are your best fodder for adding bulk and interest to an arrangement when most

Key Tasks

Cut hellebores once seed pods have appeared • Experiment with spring shrubs • Force blossom • Harvest narcissi in bud • Harvest ranunculus, tulips and weedier stems

Cut ranunculus when the flower heads are soft and squishy but not blowsy.

Narcissi are best cut when the buds are fat and in colour.

seasonal foliage is still too new and soft to cut. Various tree blossoms open in a continuous wave through spring, starting with blackthorn, wild cherries and cultivated double cherries and on to apples, pears and other fruit trees. You can, to some extent, delay or accelerate blossom's flowering time by selecting the moment to cut it. Either bring it into the dark to slow it down or place it in the light and warmth to speed up blooms by around 4–5 days. The timetable for using blossom is dictated by the seasonal variations of any given year so you do have to keep a beady eye on it to catch it at the right time.

A whole host of beautiful spring shrubs are well worth adding to your garden if you know you'll be there for years to come. We would fill our gardens with amelanchier, flowering currants, *Spiraea* 'Bridal Wreath', *Viburnum opulus* and magnolias if we had the room! Instead we make do with wandering around our site to forage from the established shrubs we can access. We've also been known to knock on the doors of people with overgrown shrubs in the local neighbourhood!

Willow provides texture in the vase with its furry buds, and all varieties root ridiculously easily when left in water so you can effortlessly add to your collection if you have space to plant it. Biennial forget-me-not grows on our plot like a weed – self-seeding voraciously – and while it isn't the most useful for stem length, it's still one of our favourites for its starry centres and vivid colour. It will lengthen and become more usable if you're patient.

It's always pleasing to cram in some biennials patiently sown the summer before (see page 156). Our wallflowers can be budding up by late winter and provide a brilliant array of jewel colours. We love 'Sunset Apricot', 'Fire King' and 'Vulcan'. Once the petals start to open they can look a little raggedy so try to cut them when there are still some flowers in bud. They start off a bit short stemmed but this improves in the second and third season. Try to cut back any spent flowers or seed heads (another great dried ingredient) to encourage the plant not to direct its energy into ripening the seeds.

We always need to grow more hellebores. The range of varieties is staggering and the little speckly star-like flowers are arguably *the* most show-stopping flower for later winter into spring. The problem with these beauties is you mustn't be tempted to cut them too early in their life cycle. Florists swear by searing the stems in boiling water, floating them in tepid water for 24 hours, slitting down the stem or other tortuous techniques. We tend to just wait until the seed pods have begun to form on the bloom and the petals have grown more leathery, then you know that the flower will be robust enough to withstand cutting without wilting.

Once spring arrives we are plunged into tulip season, heady but slightly stressful. Tulips are all about timing, watching out for signs of colour when they start to break open to know that a stem is ready to harvest. Tulips can be a bit fussy about moisture and don't like to be too wet or dry. A consistent level of water will help them develop a useful stem length but if they sit in our heavy clay soil all winter exposed to torrential downpour after downpour they may be in danger of rotting away altogether. While they have all the goodness they need stored away in the bulb, you could try a feed with liquid seaweed every three weeks or so to see if it impacts on the size of the flower head. It's not something we tend to bother with but other flower farmer friends swear by it.

Tulips are pulled, not cut, by flower farmers. This means that you maximise the stem length as you include a few centimetres of stem that were hiding underground. One of the reasons we grow our tulips in raised beds is for ease of pulling the stems rather than having to persuade them out of heavy clay soil. Aside from stem length, the other advantage to harvesting with the bulb on is that the flower will continue to be fed by the bulb while still attached – and to grow! Depending on the weather, tulips can develop and change on an hourly basis. In 2021 the blooms were developing so fast in the April heat that we were on tulip patrol three times a day, pulling up newly ready bulbs as soon as we spotted them.

Once pulled we wrap our tulips carefully in brown paper or newspaper. You are aiming to keep the stems straight but not crush the flowers. As the stems can continue to grow towards the light, be aware that if you leave them splayed out in a bucket they will assume that shape as they continue to develop. Tulips, then, should be stored in the cool and dark to hold them back until you're ready to use them. Some pro flower farmers store theirs in a chiller for up to five weeks but as we don't have access to a proper chiller (just a dark and cool room) we tend to use ours within a week.

The flowers can look a bit droopy and sad when you come to use them. To revive them, cut the bulbs off, but keep the stems wrapped in the paper so they can rehydrate with straight stems. This technique prevents those swan necks from developing, which can really limit their usefulness. We leave our tulips to complete this process in the cool and dark too so they can hydrate without additional stress. After a few hours the petals and stems should feel turgid and juicy and are ready to unwrap and use. A British, naturally grown tulip will open gloriously to its full blown potential, unlike the stick-like supermarket versions. Enjoy them in every stage of this evolution – we're not sure if any flower dies more gracefully than a tulip.

We pull tulips out with the bulb still attached for optimum stem length and storage.

A WILD MANTELPIECE

Mechanics
A tarpaulin or dust sheet
2 x reusable 60cm/24in planting troughs
Chicken wire
Some bricks or weights to place in the trough
Moss
A mix of short wide-mouthed vessels
Reel wire or thick-gauge florists' wire
2 x mossages (optional)

See page 208 for the flower varieties we used

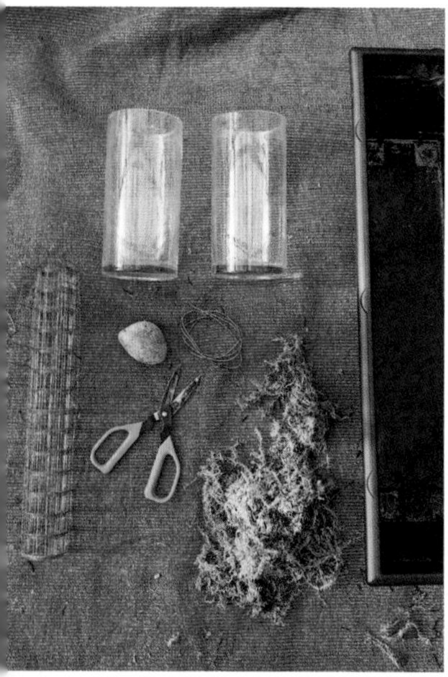

We wanted to create an abundant spring installation covering the entire mantelpiece.

For a short mantelpiece one trough would be enough but wide mantelpieces usually need two troughs side by side.

Protect the floor below your mantel with a tarp or dust sheet. Start by placing your troughs next to each other, end to end. Next take a long piece of chicken wire and wrap its length round the two troughs, but don't fasten it. Fill the bottom of the troughs with a couple of weights to keep them in position, then fill them with moss. If you plan on using thirsty umbellifers like orlaya, ammi or wild carrot they have to sit in water so you will need to nestle vessels filled with water at even intervals into the moss.

Next, pull the chicken wire over the top of the troughs and tie it to itself using wire, leaving a gap between the front fascia of the troughs and the chicken wire covering it. Fill this space with moss, allowing you to create coverage in front of the troughs for a wilder, more abundant look.

If you want to hide the troughs completely, close the ends of the chicken wire at either side of the troughs and attach two small mossages (see page 67) to each end of the troughs with wire or strong twine.

Fill the troughs with foliage first to hide the containers, then add the flowers in order of delicacy. It's important to work in this order because you risk damaging more delicate flower heads if you need to move them around. Finally, place your most delicate thirsty stems through the chicken wire into the hidden water vessels.

This is a bells and whistles kind of mantelpiece. You can create a similar effect if you have a nice-looking trough or planter by filling it with moss and covering it with a layer of chicken wire held in place with twine. If the planter is heavy enough you won't need any weights: you can nestle water-filled jam jars among the moss for stems that will need to sit in water.

SU
MM
ER

SUMMER

Summer arrives and with it comes the need to relinquish control. Although there are long and languid hours of sunlight in which to throw yourself at your list of gardening jobs, the lush growth of the plot will always triumph and get away from you. Rather than seeing this as a defeat it is perhaps more of a healthy reminder that we shouldn't try to control and constrain too much. Weeds are protecting the soil from the punishing heat; a toppled-over scabious might result in the perfect bendy stem and self-seeded grasses that pop up all over our plot and give it a rather shaggy appearance are bonuses to harvest for winter.

Embracing this lack of control is something we have always tried to carry through into our floristry. Both our weddings embraced what the season had to offer, which then dictated the colour palette for us: soft pinks, purples, vibrant greens and cornflower blues for Marianne in summer and burgundies, oranges, rusts and pinks for Camila in late autumn. The summer season in the UK offers the greatest diversity of colours and variety across the floral year – there really is no need for florists to import stems from overseas in these months. It's the time to revel in the sheer abundance and variety of flowers that we can harvest, and celebrate that the days and weeks of hard graft through the bitterly cold months have paid off. Early summer is a frantic balancing act between harvesting, planting and arranging. As the temperatures rise, there comes a moment when we have to respect the elements, succumb to the heat and work with it.

An unexpected and rewarding part of our work with brides and floral designers has been to cultivate greater awareness of the variety and beauty of British-grown flowers at the height of our season. While in summer we can dig out enough blush flowers (perhaps the colour most often asked for by brides) to cut a decent bucket, we often feel that the bride might have got the best cut from our plot if we'd let the garden lead and suggest what might be at its most beautiful in that specific, never-to-be-repeated, unique moment in time. Rather than shoehorning the garden into a particular aesthetic and hue spotted on someone else's Pinterest, we try to encourage customers to relinquish absolute control over colour and variety – much as we have to with the garden's rampant growth.

Make hay while the sun shines

Summer is, perhaps counterintuitively, the best time to start preserving and drying your own flowers to store and use through the winter. The warm air means that stems hung upside down somewhere dark with a stable temperature will dry rapidly and retain their colour. If you love the look of bleached-out stems then it is entirely possible to do this naturally in direct sunlight. We have had brilliant success bleaching Bells of Ireland, sweet pea tendrils, ammi or even bamboo foliage simply by hanging it in our glasshouse, but you could leave yours in a sunny room by the window. They are more likely to take on a yellowish hue than the brilliant white of a chemically bleached stem but are totally safe to handle versus the barium hydroxide, calcium hydroxide or aluminium sulphate (give them a Google to see the safety warnings!) applied commercially to get that whiter-than-white colour. Be aware that if left too long, the stems will become increasingly brittle, so handle with care.

SOIL

Soil in summer can feel like a world of extremes, either totally camouflaged by a thick canopy of bindweed, plant growth and annual weeds or left looking exposed, dry and unhappy when you've cleared a bed of annuals but not had something ready to take its place. The bed sits there drying out in the hot sun and rapidly tries to protect itself by spitting out a new smattering of weeds.

Keep it covered

Quick summer green manures such as mustard or phacelia are a brilliant way of getting something growing and giving a little benefit to your bed – by fixing nitrogen into the soil – while keeping your soil covered and more moisture retentive. If our heavy clay gets too dry it becomes impossible to plant anything into it. We've sown a quick green manure crop in late summer, and then cut it all down in autumn to clear space for our hardy annuals. Ideally you need to cut it before it all goes to seed.

The wonder of weeds

While it's easy to feel despairing about the terrifying rate at which weeds grow at this time of year, all this green material is excellent fodder for your compost heap which will feed the soil in the following season, and it's amazing how rapidly you can clear a huge expanse of weeds that have grown up that season. It's a supremely busy time and it's easy to simply pile all your weeds in a corner and forget about them, but try to make time to get a balance of greens and browns layered roughly throughout the heap. This is a valuable investment in your soil for next year. Remember that many ingredients in your pile may have both carbon and nitrogen in them. Decomposing ammi may have quite a woody stem (brown) but still have lots of leaves (green). Don't get overly worried about the mix – like cooking, the more you do it, the more instinctive it'll become.

We put all our perennial weeds and roots onto the compost heap, much to the horror of some allotment veterans we know. The key is to try to keep the light source reduced so you don't give them the opportunity to regrow. Try to chop up tall and bulky stalks with shears, a sharp spade, shredder or strimmer as small as you can as smaller ingredients equal faster decomposition. We cover our compost heap rather unscientifically with a tarp and just keep an eye on moisture levels so it's not too dry or too wet!

A note on biodynamic gardening

The more you compost, the more obsessive you can become about the mix and what you're adding to it. Biodynamic gardening recommends a whole host of different 'preparations' to improve your soil. There are polarised opinions about how scientifically sound biodynamic gardening is but compost queens The Land Gardeners – who do follow some biodynamic principles – recommend adding yarrow, nettles or comfrey to the heap as well as a sprinkling of garden soil or clay in fine layers throughout. Their compost cake method is a real art and well worth learning. We're not sure we have either the time, resources or belief to stuff some of these ingredients into the horn of a cow or stag's bladder as biodynamic gardening promotes, but we are yet to try it and may one day stand corrected!

Key Tasks

Sow quick-cover crops • Weed • Water wisely

Watering

On a blazing-hot day it is easy to look out at your plants and feel a wave of panic that they will all immediately curl up and die in the face of the heat. Established plants – perennials, shrubs or annuals which have been in the ground for a while and put on a decent amount of growth – will be more resilient than you think. Watering deeply a couple of times a week is infinitely better than a little nightly sprinkling. Don't just look at your soil but dig down a little and check the moisture levels below. If you regularly mulch once or twice a year you'll be surprised at how rapidly your soil is able to retain moisture and store it underground for plants to access. Remember to use those water butts you hopefully got installed in the autumn – the plants will thank you for it.

We have drip-line irrigation laid throughout our plot as we realised we were spending a cumulative total of nearly a day a week watering in our early seasons. Not only is this a massive waste of time but it's also a massive waste of water as much of it will splash onto foliage or paths only to evaporate in the sun. We plant everything near the irrigation lines to ensure the water reaches the roots effectively. The water is on a timer so it comes on in the early hours of the morning before the sun has got really hot and allows the plants to plump up before the heat of the day. Installing irrigation can be intimidating and obviously uses a lot of plastic but it should last for many seasons to come. (See page 166 for more tips on irrigation.) If irrigation feels surplus to requirements do try to water savvily and watch the weather forecasts like a hawk for forthcoming rainy days.

This page: Perennials such as feverfew will survive hot temperatures better than annual plants.

Opposite: An irrigation system keeps plants hydrated and saves water.

'Summer ushered in - how quickly the weeks fly by; the feasts of rose and lily and a hundred other sweetnesses of the garden rush upon us in an ecstasy and are gone before they can be savoured to the full.'

Simple Flowers: A Millionaire for a Few Pence, Constance Spry, 1957

SEED

In summer we reap the rewards of our early hard work. This is the season of plenty and it's time to luxuriate in the abundance of flowers in the garden. Summer is tricksy however. With one hand she fills the garden with bountiful blooms, yet with the other, pushes us away from the all-important everyday observations. Creating for summer weddings and harvesting overtakes our ability to care, tend and notice so thoroughly. It can feel like a race against time harvesting, planting out the next succession of half-hardies and keeping up with the season as the unforgiving heat emerges after the gradual warming of spring. It's vitally important to celebrate your successes when growing flowers. The failures or tasks you just didn't get round to doing can, and will, grind you down in these hot, sweaty months. Celebrate the bounty of the season as it changes from week to week. Yes, plants go over quicker as the rising temperatures accelerate their lifecycle from bloom to seed, but as one crop goes over, three others are waiting in the wings.

Despite this abundance there is still sowing to be done, some very time-specific. Biennial plants straddle two growing seasons. Sown from late spring to summer, you'll be anticipating their flowers until the following spring. These plants bridge the gap between the overwintered hardies and spring-sown crops and you'll be very grateful when they show up. A hot summer is not the easiest time to be sowing but set yourself up for success by leaving your seeds to germinate somewhere that temperatures won't swell too rapidly, and check germination regularly. Shade netting can be clipped or tied to greenhouse walls to give plants some respite. Our stalwart biennials are foxgloves, sweet rocket, sweet William, honesty and wallflowers.

Focus on foxgloves

Foxgloves are a real favourite. While all the previously harvested white froth of the cow parsley and ammi feels synonymous with late spring, early summer welcomes these statuesque dramatic spires, and the bees love them too. For those of you with shady gardens forlornly reading this book thinking that none of these flowers will bloom on your north-facing patch of green, think again. Foxgloves love some shade. Think where you've seen them growing in their true wild form. The classic purple cones of *Digitalis purpurea* are commonly found in wooded areas with dappled shade. Always remember Beth Chatto's mantra of 'right plant, right place' so that you're working with the unique conditions of your garden or allotment to grow plants that will thrive. On our extremely sunny south-facing site we built two raised beds in the summer of 2020 under the dappled shade of a plum tree for our shade-loving foxgloves and astrantia.

Patience is a must for foxglove germination. The dust-like seeds are hard to manipulate so place your hand palm up and close it slightly to create creases in your palm. Tap the seed packet so that the seeds fall into this crease and use a plant label to move the seeds from your hand to compost-filled modules. As with all tiny seeds we don't cover the seed with compost, but surface sow and give the seed tray a sprinkle of vermiculite to avoid damping off, which is caused by seeds being subjected to too much humidity and moisture. Germination takes 14-21 days, so it really is a waiting game. Once germination has

Key Tasks

Sow biennials • Begin harvesting seed for next year

occurred it's definitely worth thinning out before potting on, unless you have a lot of space for foxgloves!

Some foxgloves are reliably perennial, while others will follow the biennial life cycle. 'Sutton's Apricot' and 'Pam's Choice' are two reliable biennials, but we always sow the classic purple too as it gives arrangements a wilder, more hedgerow aesthetic and looks fantastic alongside big umbellifers. The perennial foxglove, *Digitalis lutea*, is altogether a different plant with a more delicate, cream spire of miniature bells and can be more usable in bouquets. Remember that all parts of the plant are poisonous.

Quick growers

London's balmy (at times aggressively hot) temperatures allow us a further opportunity in early summer to try direct sowing some quick growers into prepared soil (see Spring Seed, page 118) . Some of the speedier varieties we like include calendula, California poppies, sunflowers and sweet peas. This is perfect if you're not too sure about the viability of leftover seed from a previous season, as there's nothing to lose by sowing into any spare inch of soil. You'll probably find that stems will be a bit shorter with late-sown plants as the heat will encourage them to flower prematurely, so try to remember to pinch them out and remove any buds as they appear on stumpy stems to encourage the plant to branch and grow taller. Keep them well watered and feed at least every two weeks once the plants have put on a good show of growth but are yet to flower.

CAN I USE MY OLD SEED?

Growers are constantly reminded to use fresh seed and to discard old seed whose viability may have reduced over time. But for anyone growing a diverse mix of flowers, veg and herbs in a garden or allotment, this can lead to a lot of leftover or wasted seed. One way to test whether your seeds are still viable is to place them on a wet piece of kitchen paper and cover them with a sandwich bag with a few air holes pierced in it. Given time and favourable conditions, light and temperature, your viable seeds will sprout, giving you confidence to do a proper sowing elsewhere or simply pot these ones up.

Saving seed and seed sovereignty

While there's an inherent thriftiness to saving seed, there's a far more important reason to allow at least some of your crops to set seed. Seeds from a plant that has flowered and gone over in your patio or raised bed will have the right genetic characteristics to thrive in those conditions the following season.

Most flower farmers grow on a much larger scale to the humble third of an acre that we enjoy at WLFC, so allowing a bed to go to seed is something that is difficult for us to do as we are very hungry for space. Over the past two seasons, as we've expanded our patch further, we have delayed turning a bed so we can harvest our own seed and set ourselves up for success in future seasons. Perhaps leave one particularly healthy plant for the seeds to mature and ripen if a whole bed feels like more space than you can spare.

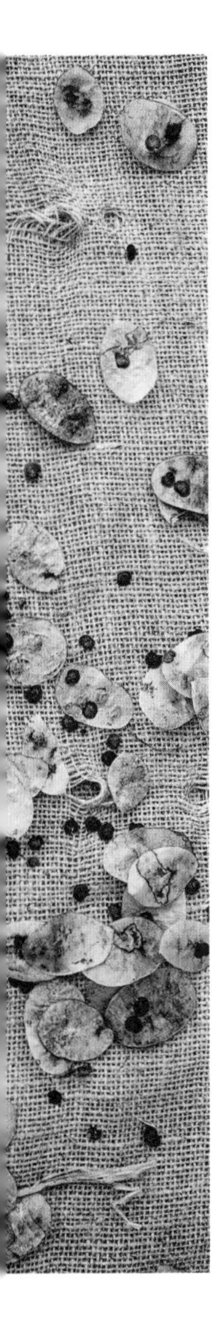

Save seed from vigorous plants with beautiful flowers for the following season.

HOW TO SAVE SEED

Late summer is the time for seed saving when lots of plants finish producing flowers and set seed. We sometimes identify a particularly beautiful version of a flower – a white scabious with a beautiful berry centre that is different to all the others for example – and try to tie a label or ribbon to this, as Constance Spry advises, to harvest the seed when it matures.

To catch it at the optimum moment, it's best to try to emulate the natural process of the plant setting seed. Allow the seed pods to mature, brown and grow crispy. A nigella or aquilegia pod, for example, will rattle tantalisingly with ripe seed when it's ready. Most will begin to split open to enable natural seed distribution to occur. Keep an eye on the seeds every few days to ensure you don't return to an empty shell where all the seeds have vanished into the soil. If this does happen you might be able to pot up the volunteers and rehome them later in the season, but they may just get lost. Harvest on a dry day as damp seed is harder to dry out and preserve. Even if it appears bone dry it's worth leaving harvested seed out on newspaper in a dry spot for a day or so to make sure there's no hint of dampness. Don't try to rush this process by placing it somewhere very warm as the viability of the seed could be compromised at a high temperature. Store your seed in a cool dark place and always, always label the date you harvested it. We store ours in money envelopes in a tupperware box.

Save seed at the optimum moment and pot up volunteers.

'In July a good many flowers ripen their seeds. If you like to make collections of home-grown seeds you will be wise to carry about with you skeins of bright coloured wools or unfadable cottons...with which you can mark the flower heads of especially good-coloured specimens. If you take care to select the finest colours of each kind of flower, you will probably have first-class seed.'

Constance Spry, *Garden Notebook*, 1940

OPEN POLLINATION

Seed saving and open-pollinated seeds are becoming increasingly discussed within the world of growing, and for good reason. Open pollination refers to the natural process of pollination by bees, birds, wind or human hands. These seeds breed true to the parent plant with a few natural modifications that set the next generation up for success. The genetic makeup of the seed is similar to the parent plant but has adapted to the unique growing conditions of its place of origin.

The ability for a plant to adapt to a habitat is fundamental when we consider how climate change and increasing temperatures affect ecosystems. Not all plants will survive, and as flower farmers we find that the summer months are sometimes a stark reminder that eventually our climate will become inhospitable for many of the plants we grow unless they're able to modify and evolve. A hybridized seed has been cross-bred to favour particular characteristics. The packet is often labelled as F1. We cannot save true seed from a hybrid plant: in many cases the seed is sterile or seed isn't produced at all so while these plants provide flower farmers with a luxury of varieties and palettes, they're not able to contribute to the garden ecosystem like open-pollinated plants do.

A diverse ecosystem is a healthy one so it's vital that we save our own seed, take part in seed swaps or buy seed from open-pollinated sources to protect floral diversity. Plants are home, habitat and a vital food source for thousands of animals and microorganisms. While some pollinators favour open-faced, daisy shapes, short-haired bumble bees have long tongues and therefore need deep flowers like foxgloves or snapdragons, so as growers maintaining a rich diversity is key. The patchwork of biodiversity created by small-scale gowers, hobby gardeners and allotmenteers across the land provides vital sanctuary and wildlife corridors for pollinators. Even your small patch can make a difference and provides an environmental alternative to the monocrops of farming on an industrial scale.

TEND

To-do before the to-do list

Summer temperatures can test even the most zen gardener's patience so set the alarm clock a little earlier in the height of summer: it will lead to fewer expletives and generally a more enjoyable day. Look after yourself as well as the plants: we put suncream on before arriving on site as we're immediately distracted by the overflowing to-do list upon reaching Wolves Lane. While we understand the appeal of wearing shorts and vest tops during the summer months, we opt instead for loose trousers to keep us cool and protected from the sun as much as possible. We also find many grasses or plants can irritate our skin and lead to rashes and grazes – yarrow is a real culprit – so keeping covered is key. A wide-brimmed hat is also essential. Yes, you will be plagued with hat hair every day, but without one you're likely to age decades quicker! While this potentially neurotic advice won't have an immediate effect on your plants, it's vitally important that gardeners protect their skin, especially given the summer heat waves that we're experiencing with more frequency.

Our most important job in the summer is to sustain our crops as long as possible. All plants ultimately want to go to seed so that they can promote the next generation of plants. When the sun is fierce, plants are liable to go to seed more quickly, using up their resources and getting more stressed by the growing conditions. A lot of summer tasks are really bound up in delaying this very process and nurturing some of the season's workhorses so that they can continue to flower.

We know that summer is upon us when we start to plant out the dahlias and half-hardies. It's imperative that you don't do this too early, and watch your local climate. The tender varieties won't appreciate a frost, however established they are, so keep checking the weather, especially the night-time temperatures, for about a week or two before planting out. We aim to get all the new dahlia varieties planted out by the beginning of June. While it's helpful to have these markers set up in the growing calendar, as always we have to work with the unique growing conditions for any season, so the schedule adapts.

Staking and deadheading

One of our most loathed jobs in the summer is staking. Leave it too late and it's an absolute swearathon as we untangle the pea and bean netting and wrestle it over plants already 70cm/28in tall. Given the ratio of stem-to-head size, sunflowers and dahlias need staking. Cosmos, even with their small delicate ruffles, have such good stem length that they too need additional support. The plants establish so quickly and grow so fast that if left unsupported they will keel over even without wind or pesky foxes. This season we're going to start experimenting with metal mesh as it's more durable and reusable. However most seasons we use wide-gauge netting, approximately 20cm/8in squared, to net the dahlias and they're held in place by hazel stakes whittled by Marianne's father so that we can write the variety on the stake itself. Once planted out it's very easy to lose small plant labels stuck futilely in the ground, and you'll inevitably trample on them as you walk up and down the beds for harvesting. We're able to recognise the dahlia varieties without the need for names, but since we don't lift our tubers every year and leave them in the ground the labelled

Key Tasks

Stake • Deadhead • Tidy paths and re-establish plot borders • Pack dried flowers away for winter • Watch and manage pests

stakes show us where a tuber hasn't survived the winter and a gap needs to be plugged. With a plot as small as ours, it's imperative that we make use of all available space.

While early spring is full of anticipation and an almost desperate need for the plants to start flowering, summer holds different anxieties. Warm growing conditions mean that the varieties that make up the lion's share of our summer crop start to flower at such a rate that we can't always cut quick enough. We allow pollinators and other wildlife to enjoy the blooms we don't cut but deadheading is a must if we want the plants to continue flowering into late summer. When deadheading dahlias, we cut the stems right down as this will promote longer stem length for future blooms. It's always worth finding time to deadhead cosmos, sunflowers, roses, nicotiana, ammi, wild carrot, French marigolds and Mexican sunflowers, *Tithonia diversifolia*.

You might imagine that our houses are full of blooms throughout the season but we very rarely make time to cut for ourselves, which is a failure on our part. It's not only a treat to keep flowers for ourselves, it's also a way of getting better acquainted with new varieties, making sure that we're cutting at the optimum point and observing how they age. Some flowers just get better and better until they disintegrate altogether – tulips and roses are prime examples – but it's impossible to know this if every stem is cut for someone else. It's also important to keep joy firmly at the centre of any growing project, so do cut yourself a handful of stems at the end of every week.

Maintenance

Although we associate this time of year with seasonal abundance, midsummer can be surprisingly lean as the late spring and early summer workhorses go over and roses have finished their first flush. Look after your summer stalwarts by treating cut-and-come-again crops to an organic seaweed feed every couple of weeks so that the plants don't become depleted.

Once the biennials have been sown and you're keeping up with the deadheading it can feel like there's a slight reprieve from the punishing spring growing schedule. But look up and you'll find the bindweed has gone to town on all the stakes and compost bins and other garden thugs have started popping up where gaps have appeared in the wood-chip paths.

Midsummer is the time to tackle these issues one step at a time. If, like us, you've had your head down and focused on the actual growing of plants for the best part of three months you probably haven't had much time for tidying and maintenance, so the garden might be looking pretty wild with grasses, weeds and self-seeders. Take heart: even the most experienced gardeners can suffer from gardening overwhelm, and it can feel particularly acute during the summer months when the garden firmly reasserts that it is the one in control. Even if you only have five minutes, don't just stare in dismay – like we have on many occasions – but fill a wheelbarrow with wood chip and address any paths that need help. Weeding may not be the best use of time, especially since prevention is so much more efficient than cure, but if your plot is starting to look more like an art installation than somewhere crops are grown, weeding can provide much-needed therapy and a surprising tension outlet.

Irrigation

If you've invested in an irrigation system, tweak the timings if necessary. Plants should be irrigated either early in the morning or late at night and summer temperatures might mean that the system needs to run longer. Do carry out regular checks and replace the batteries in the timers before they run out. Since many of us aren't in the garden or even awake when the irrigation comes on, it's hard to know if there's an issue until it's too late. Stick your fingers in the soil regularly to check for moisture levels and if the soil feels dry, there's a problem. We carry out maintenance checks by switching the timers on for a couple of minutes and walk around listening to the pipes pressurize and start to drip. We know we have a leak if we can hear hissing, which means water is escaping somewhere. Leaks on our site are most commonly caused by young foxes using the drip line as a teething toy so we keep a store of spare drip pipe to replace any sections that the cubs have chewed through.

For those of you without an irrigation system make sure you use water and your time as efficiently as possible. Save yourself and the leaves a scorching by watering as early or late as possible and give your plants a long soak once or twice a week rather than a short watering every day. This encourages roots to dive deeper down into the soil bed for moisture rather than establishing shallow growth near the surface.

Drying racks

The summer allows what we refer to as 'the drying station' – essentially just a repurposed staging grid which we hang stems off – to hit peak productivity. Keep an eye on your drying area and wrap and box anything that has dried to make room for new crops. They come thick and fast in summer and although your mind might be focused on the abundant fresh crops, you'll be pleased you squirrelled away surplus for winter projects come the end of the season. Store your boxed cache somewhere dark, cool and dry to preserve colour and to avoid heat desiccating stems and making them overly brittle. We leave a few crops to bleach naturally, notably ammi, sweet pea tendrils and bells of Ireland.

Wrap and box all stems once dried to make room for more flowers.

COPING WITH PESTS

Self-seeded plants tend to be tougher and more resistant to pests than the ones you lovingly sow and cosset from autumn through to spring. We love to grow ammi in the glasshouse and plant it out in early winter to be rewarded in late spring with a towering mass of stems, but this sheltered environment also ushers in thousands or hundreds of thousands of aphids and spider mites in every variety imaginable. Growing without pesticides means the battle against these pests is inevitably lost – in 2021 we had to pull out two destroyed beds' worth of ammi before any of the flowers had fully matured.

Pests will simply migrate to another similar plant if they're disturbed so work in a careful and slow fashion when removing infected plants. Violent outbursts towards aphids are understandable but unfortunately not that effective. Our best strategies against pests are to keep plants healthy with consistent watering and seaweed feeding, to remove any signs of pests by hand when first spotted, or to be ruthless and cut out whole stems or plants where an infestation is concentrated. If necessary we also buy in ladybird or lacewing larvae (or bring any we spot outside into the glasshouse) from May to July to help keep on top of the problem. This won't be effective if you're growing solely outdoors as it's difficult to keep the larvae where you want them. If you have a healthy ecosystem with plenty of diversity, no chemicals and healthy soil, hopefully the ladybirds will come to you.

Even the most seasoned growers will regularly experience failure. Umbellifers like wild carrot can be felled by carrot root fly, low-flying insects that land and lay their eggs in the surrounding soil where the larvae hatch in spring and happily munch through the roots. You'll be blissfully unaware until the foliage begins to curl and crisp and the plant eventually keels over. Netting around the plants with a fine mesh that's at least 20cm/8in tall to prevent them getting to the plants or using a specially designed nematode you can buy online are worth a try, but we're still experimenting.

Generally you need to learn to accept a degree of loss from pest damage. The best approach is prevention not cure. Healthy, robust well-established plants grown in healthy soil will withstand night-time raids from slugs far better than a spindly seedling. The inclusion of a pond on our plot, we hope, will pay dividends in the future as frogs and toads move in and hopefully thrive on the population of molluscs.

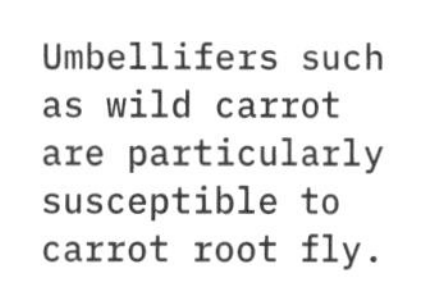

Umbellifers such as wild carrot are particularly susceptible to carrot root fly.

HARVEST

Harvest fresh and dried flowers

Summer's harvest is unrivalled. The kaleidoscopic diversity of blooms throughout the season should inspire any flower lover to try growing their own, even if just starting in a window box. Jotting down an exhaustive list of flowers we harvest during the summer months feels overwhelming (even to us!) so keep it simple with three to six varieties of annuals if you're new to all this. Some will be suited to sun or shade so find out as much as you can about what the plant needs to thrive before getting seduced by the entire seed catalogue. Some of our favourites are listed over the page.

Our primary focus here is annuals, perennials and biennials but if we had our time again we'd invest more time and money from the outset into shrubs and perennials. They take longer to establish but they can really give nuance and variety to your plot, and often withstand drought or drizzle better than short-lived annuals. Cut sparingly or not at all in the first season of establishing new perennial plants so the plant can concentrate on putting its energy into establishing roots and healthy growth. Do deadhead religiously to prevent the plant redirecting its energy into setting seed too. You will be rewarded the next season with far more flowers if you allow the plant the first year off.

How and when to harvest some of our summer stalwarts

In the height of summer it's critical to harvest flowers first thing (before you can feel the sun warming the back of your head) or failing that, at dusk. We get very nervy when customers want to collect flowers from us in the middle of the day during summer as this is a sure way to stress the flowers and reduce their vase life. Flowers should be cut into deep buckets of clean, fresh water and left somewhere cool and dark to rest and rehydrate for at least half a day, ideally a full 24 hours before using. Keeping your buckets clean is a tedious but vital job to ensure the very best vase life.

Harvest flowers first thing in the morning or at the very end of the day.

Key Tasks

Harvest flowers first thing in the morning, or in the evening • Cut flowers for drying • Deadhead cut-and-come-again crops

Some of our favourites.
All are good for cutting.

Perennials

Achillea
Astrantia
Catmint (*Nepeta*)
Dahlias
Delphinium
Echinops
Eryngium
Fennel
Feverfew
Gladioli
Goat's rue (*Galega officinalis*)
Lady's mantle (*Alchemilla mollis*)
Lupins
Peony
Perennial phlox
Roses
Salvias
Sanguisorba
Scabious
Verbascum

Annuals and biennials

Allium
Amaranth
Ammi majus, *Ammi visnaga*
Annual phlox
Basil
Bupleurum
Calendula
Chamomile
China aster
Chinese forget-me-not (*Cynoglossum firmament*)
Clary sage
Corncockle
Cornflower
Cosmos
Edible chrysanthemum
Florist's dill
Foxglove
Gladioli
Larkspur
Malope
Nicotiana
Nigella
Orlaya
Ornamental grasses
Poppies
Scabious
Snapdragon
Statice
Strawflower
Sweet rocket
Tithonia
Velvet trumpet flower
Wild carrot
Zinnias

Gladioli

Alchemilla mollis

Poppy

Feverfew

Calendula

Snapdragon

Larkspur

Malope

Basil

Sunflower

Zinnia

Cosmos

Orlaya
Dahlia
Fennel
Strawflower
China aster
Sanguisorba
Rose
Annual phlox
Nicotiana
Velvet trumpet flower
Corncockle
Delphinium
Bupleurum
Nigella
Foxglove
Amaranth

FAVOURITES FOR WEDDING BOUQUETS

Umbellifers

Blooms like wild carrot, orlaya, ammi, wild fennel and florist's dill should have a leading flower head that is fully open, like the palm of your hand carrying a tray of drinks. Snipping them prematurely will result in droopy stems that can't be revived. They'll sulk, too, if you harvest them in the blazing sunshine and must be cut straight into water. If, occasionally, you are so pinched for time that you have to cut in the middle of the day, make sure to recut the stems into deep water and place in a dark cool room. If left overnight they should revive.

Annual phlox

'Crème Brûlée' and 'Cherry Caramel' are exquisite additions to a bride's bouquet. They're among the few flowers that genuinely fit in the 'nude' or sludgy brown palette that is often requested by brides and florists. There are lilac and pink hues among their sandy coloured petals and they are beautiful. Initially the plants flower on extremely short stems in a sprawling, rather untidy fashion but keep cutting them hard to encourage longer stems. Plant them much closer together than you think too. In 2021 we tried a 20cm/8in spacing and could have gone tighter.

Cynoglossum

Cynoglossum amabile, Chinese forget-me-not, provides a more refined version of our wild forget-me-not natives and a welcome pop of blue in a bouquet that is increasingly hard to find as the season progresses. Dainty little flowers run the length of the silvery green stems but wait until all the flowers have opened up on a stem before harvesting. If impatience gets the better of you, the stems will droop and are difficult to revive. We sow this in the autumn but they will also do fine from a spring sowing. Keep cutting the stems and the plant will continue to produce its delicate flowers, and when the plant gracefully goes over, the seed heads are a wonderfully structural dried ingredient too.

Annual phlox is a favourite for summer bridal bouquets.

SHADE LOVERS

Astrantia

East- or north-facing garden owners rejoice, astrantia is keen on partial shade and will continue to flower throughout the summer. It is notoriously difficult to get started from seed but self-seeds vigorously so look out for volunteers, dig them up and pot them on as they establish, to be planted out in the autumn. Cut astrantia once the lead flower head is fully open. White forms can get rain damaged and brown quite easily. For that reason we really like the deep pink to red varieties.

Alchemilla mollis

Another perennial that will thrive in a shady corner and can sometimes flush twice in a season. Wait until the flowers have matured but try to catch them when they're still at their most acid green. The stems dry beautifully too and are a useful frothy filler for winter arrangements.

Sedum

Not the sexiest of flowers, but a total workhorse from late summer through to autumn as a filler-foliage in arrangements and bouquets. If cut when the green is just flushing with a hint of pink it can provide a welcome pop of the elusive 'blush' that is so often requested. We prefer to cut it when it's in bud before the flower becomes huge, flat and rather unusable. Any flowers that don't get cut will dry beautifully as wonderful sculptural seed heads for winter arrangements, or leave the stems on the plant for a statuesque silhouette over the cold months – insects will be grateful for the hollow stems to overwinter in.

BEST FOR POLLINATORS

Fennel

We cut fennel at all stages of flowering, from the first flush of vivid yellow right until it's gone fully to seed. It's a vital summer umbellifer from mid-season on when others may be fading. A voracious self-seeder, it can be a thug but we're always grateful for the natural succession of blooms this produces, sometimes as late as midwinter. Pollinators love fennel so don't forget to leave some heads for the hoverflies and bees.

Cut fennel at all stages.

Nepeta

Nepeta is always abuzz with pollinators, and it's commonly called catmint as it can create a frenzied drug-induced high among cats. Plants may not thrive if you share the garden with a large feline population, although inserting a few short stakes to stop them trampling the crop can help. After flowering, cut plants right back for a second flush, and even when the flower heads have begun to fade the lavender-coloured seed bracts can look beautiful as a filler. *Nepeta* 'Six Hills Giant' is our favourite variety for stem length but it does sprawl so needs staking or netting.

Foxgloves

We always end up with more foxglove plants than we need but are grateful for their tall spires to add long-lasting drama to an arrangement. We plant our foxglove glut in several areas from deep shade to full sun to stagger when they bloom. As they age you can still harvest them by removing spent blooms from the bottom of each stem, and once the first, showiest blooms have finished, shorter side shoots will appear that are often more useful than the tallest stems. Although sown in midsummer as a biennial, they can often return as a short-lived perennial.

Catmint is a summer stalwart and always buzzing with pollinators.

PROLIFIC FLOWERERS

Dahlias

There are thousands of different dahlia varieties that have become increasingly sophisticated in palette and shape thanks to American flower growers who have started breeding their own varieties with the wedding industry in mind. Don't get too hung up on the different classifications (Ball, Pompom, Decorative, Waterlily etc) just for the sake of knowing them, but different types do have different optimum times for cutting.

Florists can be sceptical about dahlias. Market-bought stems can go over very quickly but this is not the case with locally grown dahlias. If they are cut at the right moment, dahlias happily last five days. Once cut, they won't keep opening – cutting them stops them in their tracks – so if you're planning on harvesting some blooms it's important to do it at the right moment.

As a general rule you want the flower to have developed its form but still have a flat back before it starts reflexing over the stem. For single varieties – we love 'Honka Fragile', 'Verrone's Obsidian' and 'Waltzing Mathilda' – you're looking for the centre to be a tight pincushion of balls before the bees have arrived and fluffed up the pollen. Once pollinated, the blooms won't last more than a day or so but we still maintain they are an essential addition to the cutting patch for their beautiful form in an arrangement.

Waterlily and decorative dinner-plate dahlias like the feted 'Cafe au Lait' need to be approaching total 180 degree openness. They can be cut when they're just shy of this moment – the bud just bursting open – to make them more workable in an arrangement, and we prefer to cut them at this point otherwise they can appear to stand to attention without a lot of movement.

Dahlias love and need to be cut to prolong their flowering period and you should be bold and reach deep into the plant to cut a long stem to encourage the growth to regenerate from low down in the plant. If you're planning to harvest a lot of dahlias on a specific date, be ruthless with your deadheading roughly five days before to give them time to regenerate their blooms.

Zinnias

Zinnias are managing to shake off their reputation for being slightly bonkers and seriously unfashionable flowers, thanks again

primarily to American breeding programmes. Zinnias germinate rapidly but are delectable to slugs and snails. We're lucky to be able to grow most of ours in the glasshouse (too hot for most of the slimy pests) where they can grow up to 2m/6ft tall. If you're only able to grow outside try to grow them on somewhere you can keep an eye on them until they reach sturdy plants before putting them in their final position. This can help to prevent the worst of the slug damage. They'll also benefit from a pinch (see page 129).

We always follow Erin Benzakein of Floret's 'wiggle' test before cutting to ensure they're ready to harvest: hold the zinnia stem about 20cm/8in below the flower head and give it a gentle shake. If the stem remains rigid with the flower moving from side to side it's good to cut. If it still feels pliable it's not quite ready yet and is liable to snap once harvested. Also check whether the pollen is fluffing up from pollination as it'll then start to fade more rapidly.

Sunflowers

The floristry world is awash with snobbery about sunflowers because of their unfashionable bright yellow petals, but you're missing a trick if you don't consider the wonderful varieties 'Vanilla Ice' or 'Pro-cut Plum'. Easy from seed, sunflowers tower above the beds and inspire awe from the youngest to the oldest of flower growers. Branching varieties give you the most bang for your buck – you can keep cutting and cutting them to encourage more stems. Sunflowers open rapidly in the sun so cut when the petals look fully formed in bud but are not yet opening up.

Sweet peas

The ultimate cut-and-come-again flower. Sweet peas can be cut from late spring if overwintered or right through into late summer if sown in spring and then cut every few days. Sweet peas are extremely hungry plants with long roots so will need feeding every week and they will not tolerate dry roots. The first flush of blooms will reward you with long stems but these shorten as the plant ages and temperatures ramp up. Either use the shorter stems in bud vases or keep on top of the rampant growth by harvesting whole tendrils at a time with multiple flowers branching off the stem. Sweet peas have a fleeting vase life of around three days but they're so worth it for the scent! The strongest scented varieties last for the shortest time in the vase so choose which seeds to sow carefully depending on your preference for longevity or scent.

Above left: Cut dahlias when the flower head has a flat back, before it starts reflexing over the stem.

Above right: Sweet peas are the ultimate cut-and-come-again crop.

BEST FOR DRYING

As farmer florists we take great pains to learn the best time for cutting flowers so they last longest. No farmer can tolerate waste because the harvested crop has to make the initial upfront investment of seeds, compost, equipment and time worthwhile. However, if you have a glut of a particular crop it's always worth experimenting with drying. We dry virtually everything by hanging it upside down in the dark. Exactly when certain plants need to be harvested for best results takes some trial and error, but on the whole we find many flowers like to be at the point just before they're at their peak blowsiness to dry well. In general, if stems have matured too far they risk falling apart; not open enough means they'll shrivel and have less impact. Experiment and you'll be amazed at what brilliant results an array of flowers can give.

Strawflower

Slightly despised for several decades as a bit of a 1970s throwback, we find cutting our strawflowers before they have fully opened is a good way of toning down their brash looks before the bright yellow yolky centre appears. They will keep opening a little once cut, so the right moment depends how you like them to look. Once hung to dry, the stems get very brittle so may need to be attached to skinny, sturdy wire to be incorporated into wreaths or nestled very deep to be supported by the other stems. If carefully stored they can last for years.

Statice, strawflower and ornamental grasses such as 'Bunny Tails' are perfect for drying.

Statice

We focus on the apricot, white and pink shades but statice comes in blues, purples and yellow too. Wait until the papery blooms are fully open before harvesting, and keep cutting for a profuse repeat flowering. We think these are seriously underrated flowers, and they make particularly brilliant additions to buttonholes or flower crowns that need robust blooms that won't sag and sulk during a day of festivities.

Everything else

There are obvious candidates, including China asters and gomphrena, but don't ignore others: larkspur, feverfew, cornflower, tansy and achillea offer wonderful dried textures. Delicate blooms like larkspur dry best when a few buds are still intact at the top; others dry better if harvested fully open. The key is to experiment. We've had surprise successes with rocket flowers, wallflowers and ranunculus. Keep a wide sheet of paper underneath your drying stems to catch any falling petals for confetti or seeds for next year.

Roses

Garden roses are commonly and incorrectly accused of being short-lived in a vase. While they won't last for over a week like some of the tightly budded supermarket offerings, the secret is to harvest them at the right time and celebrate and enjoy them for the gloriously ephemeral bloom that they are. A rose that looks glorious on the plant has already passed its best for cutting and its acceleration towards death is inevitable. For best results cut when the sepals have come away from the bud, folding back towards the stem but the flower is yet to open. If cut on a Thursday into deep, cool water and a clean vessel our roses are then at their best for a Saturday wedding and will last a day or two more if properly looked after. You can speed up and slow down the process to an extent by bringing the flower in or out of direct light, somewhere warmer or cooler but each rose is its own flower and there are no exacting rules. Like all gardening, you're constantly undergoing a series of miniature science experiments where the variables of when you cut it, variety, point in the season, health of the plant, vessel and quality of water you're cutting into render the pursuit of total control futile. You can only observe and adapt the conditions for the rose as you see fit.

Which roses are good as cut flowers? Everyone starting off in flower farming views the answer to this question as the holy grail. To some extent you can cut any rose and enjoy it in a vase for a time. How long you want it to last, how it will be used, where it will be displayed and what you like are all key considerations, and if that feels maddeningly evasive we've listed a few varieties we love on the next page.

Admittedly some roses are better for cutting than others – single and rugosa rose varieties will always be more fleeting than their shrub and floribunda counterparts. Don't get too hung up on rose classification – there are hundreds of varieties and you can cut any, but avoid the heavily thorned ones – we love 'Wild Edric', a hot pink shrub rose with masses of flowers, but the thorns will shred your hands without a pair of suitable gloves. Single varieties with open centres will have a shorter vase life so you might opt to leave these primarily for the bees.

Many garden roses, such as 'Lichfield Angel', 'Lady of Shalott' and 'Boscobel' are perfect for cutting.

OUR TRIED AND TESTED FAVOURITES FOR CUTTING

- 'Grace' – a strong apricot colour with quite a fancy centre of double petals, a favourite for bridal bouquets.
- 'Hot Chocolate' – a beautiful, saturated shade of red which is a prolific flowerer and will make you rethink your dislike of red roses.
- 'Just Joey' – long-lasting apricot blooms which open into quite a flamboyant ruffled shape on long stems.
- 'Koko Loko' – not the most vigorous of flowerers for us but well worth it for the smudgy, brown, pink and blush shades and elegant petal shapes. Brides always want a bucketful and we can only cut a handful!
- 'Lady of Shalott' – a prolific, vigorous apricot shrub rose which flushes pink in bud.
- 'Lichfield Angel' – pale peachy pink, practically thornless with excellent long stems.
- 'Queen of Sweden' – beautiful cup form with excellent vase life, although a bit of a saccharine pink.

When selecting your roses, scent should be all but non-negotiable. The imported roses that make it into our supermarkets smell of nothing as this characteristic has been bred out to prioritise longevity. Home-grown alternatives are bendy stemmed, generous and smell amazing. It's really important for us that we keep scent at the forefront of our minds when we're making bouquets. It's one of our most powerful senses for evoking key memories and that rose can become forever tied to our wedding day, the departure of a loved one or even an overarching memory of childhood. Scent is essential for biodiversity and for humans, it uplifts us from our concrete jungles and plonks us straight back into the midst of nature.

WHY GROW ROSES WHEN THEY'RE SO READILY AVAILABLE AT THE SUPERMARKET?

For wildlife

Chances are, most readers of this book are keen to include more flowers to cut from their garden but aren't planning a 15-acre flower farm turned over to roses. If this is the case, keep roses with open centres and easy access to pollen at the top of your list. If cut early as described, you'll be able to enjoy these for three to five days on the kitchen table and leave the majority of the flowers for the insects to feed on and benefit from.

For the environment

The flower section at the supermarket provides roses all year round, but they're actually only in season in the UK from May to September or October. Cut roses make up an estimated 30 to 50 percent of the €20–25 billion retail value of cut flowers overall in Europe, so that's a lot of roses being grown in heated glasshouses in the Netherlands or being flown across the world from Ecuador, Ethiopia, Kenya or Colombia. We can enjoy our own glorious, scented, pesticide-free, locally grown roses for at least five out of twelve months of the year. That's not bad. Living seasonally gives us more awareness and enjoyment of flowers as they start to bloom. We understand the desire to have roses available year round but living without these stems for over half the year means that we appreciate them tenfold as spring slips into summer.

Roses are definitely not the cheapest to grow but they will grow in pots on a patio if kept well fed and watered. While at WLFC we're purists when it comes to seasonality, we advise that if you must buy out of season flowers from abroad, then please consider buying Fairtrade flowers (see page 73).

TABLE RUNNER

Mechanics
Lengths of copper gutter
Moss
Small test tubes or glass vials
Copper mesh
Copper reel wire or twine

See page 208 for the flower varieties we used

Start by filling the gutter with moss, nestling tubes in it and filling them with water. Then cut a strip of copper mesh the same length and width as the gutter and lay this over the top of the moss. Run a length of copper reel wire or twine lengthwise over the mesh insert and under the gutter, and tie it off tightly at one end, then start to place your flowers, making sure to use the vials for umbellifers and other stems that like to sit in water. There should be space between the flowers to maintain a sense of airiness. Ensure you keep an eye on water levels and replenish if making ahead of an event as stems will drink water quickly.

Hide any plastic vials by covering them with additional moss at the end or leave more attractive glass vials proud of the arrangement to make them a feature.

CALENDAR

It is notoriously difficult to pin exact dates to jobs, and seasons can meld into each other; the tasks for each crop move a little to align with the unique weather conditions. We hope this simple calendar won't overwhelm new growers. It is not exhaustive, but should be useful to refer to throughout the year.

	SEED SOWING	BULBS
EARLY AUTUMN	Sow hardy annuals and perennials	
MID-AUTUMN	Sow hardy annuals	Begin planting narcissi, fritillaria, alliums
LATE AUTUMN / EARLY WINTER	Sow sweet peas	Plant out tulips
MIDWINTER	Do a seed inventory and order replacements	
LATE WINTER / EARLY SPRING	Sow hardy annuals under cover	Keep an eye on rain levels with tulips to safeguard stem length
MID-SPRING	Sow half-hardies under glass Direct sow hardy annuals	Begin harvesting
LATE SPRING / EARLY SUMMER	Sow biennials and perennials Direct sow fast-flowering hardies and half-hardies	Allow spent foliage to die back Cover beds or plant annuals on top
MIDSUMMER		Order next season's bulbs
LATE SUMMER	Sow hardy annuals	

DAHLIAS	CORMS	PERENNIALS
Harvest. Evalulate which plants you want to keep next year, where you're missing particular colours or shapes, and make a note of others on your wishlist using Instagram		
Harvest	Pre-sprout anemone and ranunculus corms	Divide and mulch
Cut down dahlias after first frost. Lift tubers or mulch if leaving in the ground	Harden off and plant out sprouted corms	Plant bare-root roses or trees
Order new tubers and cuttings for the season ahead		Plant bare-root roses or trees
Pot up stored or new dahlia tubers in a protected greenhouse	Start a final succession of corms Harvest anemones	Plant new perennials
Take dahlia cuttings	Harvest autumn-planted ranunculus	
Plant out dahlias when the risk of frost has passed Feed dahlias as they begin to bud up Harvest begins	Harvest	Cut back spent perennials for a second flush where applicable
Ensure all your dahlias are comprehensively labelled for easy identification in autumn	Lift spent corms when foliage has died back and store	
Keep on top of watering and deadheading	Order corms for following season	Cut back spent foliage and seed heads

GLOSSARY

BACTERIA
Bacteria are one-celled organisms generally 0.004mm wide. A teaspoon of productive soil contains between 100 million and 1 billion bacteria. Bacteria reproduce quickly when optimal conditions occur in the soil. We often think of bacteria as being something to eliminate, but they are vital microorganisms in healthy soil. For example, they process atmospheric nitrogen into a form plants can use.

BIENNIAL
A plant that is sown in spring and summer and flowers at around the same time the following year. Biennials include foxgloves, forget-me-nots, honesty, sweet rocket, sweet William and wallflowers.

BIODYNAMIC
An approach to gardening pioneered by Rudolph Steiner in the 1920s. The practice aims to incorporate science and the rhythms of nature to grow the healthiest plants, and to identify the optimal times of the month to undertake gardening tasks such as sowing seeds. Biodynamic gardening places a great emphasis on soil health and uses specific 'preparations' to add particular ingredients to the soil, compost heap or as a foliar feed for plants.

BOKASHI
A composting process from Japan, most commonly used with food waste, that uses fermentation to anaerobically decompose waste into organic matter. Bokashi bran can be purchased alongside a couple of airtight tubs so you can experiment with this process at home. It is particularly suitable if you are short on space for composting.

CARBON FOOTPRINT
A carbon footprint is the best estimate available of the impact an entity has on climate change. This can include direct energy use such as transportation but may also include indirect emissions in relation to its production too.

COIR
A densely compressed fibrous material derived from coconut waste. Coir is a low-nutrient growing medium, particularly useful as a peat replacement in seed-sowing compost. It is shipped from countries such as Sri Lanka, Indonesia or the Caribbean. Always purchase from a reputable supplier.

COLD STRATIFICATION
The process of subjecting seeds to a period of cold, winter-like conditions to mimic their preferred germination conditions. Seeds such as larkspur can be placed in the fridge for a couple of weeks before being sown to encourage better germination. If you have had poor germination from fresh seed it may well be worth trying this process.

COMPACTION
A term applied to soil that has had its structure compromised or crushed by too much weight from people, machinery or building work. Healthy soil should have a porous, honeycomb-like structure to enable it to remain aerated, for water to flow through it easily and to encourage beneficial microorganisms.

COVER CROP
A variety of plants that can be grown to protect soil from rain, wind or sun exposure. Many have nitrogen-fixing properties that bring nitrogen from the atmosphere down into the soil. Common examples of useful cover crops include clover, phacelia, mustard, buckwheat and vetch.

CUTTINGS
A form of propagation suitable for many flowering plants, using pieces of stem with buds and large leaves removed to create new plants. A cutting is placed into free-draining soil so that roots can develop before any rot occurs.

DAMPING OFF
A fungal disease that infects newly germinated seedlings and causes them to collapse and die. Commonly caused by poor ventilation, too many seeds grown too densely together and too damp or humid conditions, and very common in greenhouses. We find using a top layer of vermiculite on our seed trays helps prevent damping off.

DEADHEAD
Removing spent blooms from a plant to encourage new growth and to prevent the plant from prematurely setting seed. Always deadhead back to a point of new growth or a join in the plant's stem rather than simply snipping off the bloom. This will encourage growth to regenerate from this point.

DIVISION
A method of propagating from an existing perennial, rhizome or tuber. After several years of growth, many plants lose some of their vigour and the centre of the plant becomes woody. By digging up and slicing the plant into new smaller plants you can kickstart the plant back into productivity.

DRILL
A straight, shallow trench made in the soil in which to sow seeds directly. Sowing in a row enables you to identify seedlings over weeds more easily.

F1 HYBRID
An abbreviation of 'Filial 1' referring to the first generation of offspring of distinctly different parent plants. The best qualities of two plants are selected and bred into the new hybrid, a process that can take years to perfect. F1 hybrids will not come true from seed.

FAIRTRADE FLOWERS

The only consumer-facing accreditation. Fairtrade certified farms have to meet certain standards in terms of safety, working conditions and fair pay. Fairtrade farms receive a premium of 10% on every stem to invest in workers' healthcare, education, gardens for them to be able to grow their own food or other benefits. Fairtrade flower farms are primarily in African countries such as Kenya, Ethiopia, Uganda and South American countries as well as India.

FLORAL FOAM

A lightweight, porous substance invented in the 1950s, initially considered a miracle substance that could keep flowers hydrated without directly placing them into water for several days. Floral foam is not biodegradable, not recyclable and contains formaldehyde, which is toxic to humans, animals and marine life. Chicken wire, moss or floral frogs are all viable alternatives to floral foam.

FLORAL FROG

Also known as a Kenzan. A heavy, metal spiked base that can be placed at the bottom of a vessel and into which you insert flower stems, creating structure and shape within an arrangement.

FOLIAR FEED

Fertiliser in the form of a compost tea, liquid seaweed or weed juice that is sprayed directly onto the foliage of a plant to improve vigour.

FUNGI

Fungi along with bacteria are vital microorganisms that aid decomposition in the soil. Fungi are microscopic plant-like cells that grow in long threadlike structures or 'hyphae' that make a mass called mycelium. Fungi convert hard-to-digest organic matter into a usable form for plants. Hyphae help bind soil particles together, creating a structure that helps keep soil oxygenated and increases its water retention.

GERMINATION

The transformation from seed into shoot. Germination occurs when there are favourable conditions of light, moisture, temperature and season for the seed to grow.

HARDENING OFF

Plants that have germinated and established under cover in a polytunnel or greenhouse need to acclimatise before being planted outside. Hardening off is the process of getting the plants used to cooler temperatures, wind and fluctuations in moisture. We harden off our plants over five days in a sheltered spot where we can keep an eye on them so that they don't dry out. Others recommend a longer hardening off period, but we rarely have the time for this.

HALF-HARDY ANNUAL

Tender annual plants that require mild, frost-free temperatures to germinate and thrive. Like hardy annuals, half hardies will germinate, mature to a full plant, flower and set seed within one year. Typically, half-hardies are sown in late spring through to midsummer and finish flowering productively by mid-autumn, depending on temperatures.

HARDY ANNUAL

Hardy annuals are sown, grown, harvested and set seed over the course of a year. Many annuals like larkspur, nigella or orlaya like a period of cold stratification and perform their best when sown in the autumn and overwintered either in the ground or pots, to flower the following late spring or summer. Overwintered hardies can withstand temperatures to around -3°C. Late winter to early summer is the second period of the year when hardy annuals can be sown. These seeds will mature, flower and set seed in that same year.

HOT COMPOSTING

Ingredients are carefully layered within a compost heap to promote temperatures of 40-60°C, where microorganisms can thrive and break down organic matter rapidly. This can happen within 90 days. A hot composter is essentially a heat-retaining insulated box, ideal for food scraps as it is vermin-proof and can process cooked food including meat and bones.

HUMUS

Organic matter within and on top of the soil.

MICROORGANISMS

Also known as microbes, the millions of bacteria, fungi, protozoa and nematodes that inhabit the soil. There are more microbes in one teaspoon of healthy soil than humans living on the planet.

MOSSAGE

An alternative to floral foam. Damp moss is encased in a chicken wire pillow, secured by twisting the wire onto itself or using additional wire if necessary. Straw can also be used as a filling if using robust stems that need minimal hydration.

MYCORRHIZA OR MYCORRHIZAL FUNGI

Beneficial fungi grow around plant roots, taking sugars from plants in exchange for moisture and nutrients gathered from the soil by the fungal strands. The mycorrhiza greatly increases the absorptive area of a plant, acting as an extension to the root system.

MYPEX

Light-suppressing black plastic membrane that allows rain to penetrate, often used to cover soil or pathways to stop weed growth. Please note that it does not biodegrade.

NEMATODE

Nematodes are non-segmented worms typically 0.05mm –1.27mm in length. 95% percent of nematode species are beneficial to the soil; only 5% cause problems. Nematodes eat bacteria, fungi, protozoa or other nematodes depending on their subspecies. They play an important role in releasing minerals and nutrients in a form that plants can access.

NITROGEN FIXING

Nitrogen-fixing plants such as clover or lupins convert nitrogen from the atmosphere into a soluble form accessible by plants. Bacteria clustered in nodules around the roots of these leguminous crops convert the nitrogen into a form plants can use. Once the plants die or are cut back this nitrogen stays in the soil to benefit the next crop.

NO DIG

A method of growing that involves disturbing the soil structure as little as possible. Weeds are initially covered with cardboard then layers of organic matter such as compost or manure are added on top. Weeds are deprived of light and struggle to resurface compared to those in a bed which has been dug over, where weed seeds have been brought to the surface.

NODE

A growth point on a stem or a knobbly joint from which roots or new growth can shoot, when taking a cutting.

NPK

The three elements needed by plants to thrive: Nitrogen (N), Phosphorus (P) and Potassium (K). Nitrogen helps plants to put on green growth, phosphorus helps with fruiting and flowering and potassium with developing root systems and protecting plants against diseases. Different plants require different ratios of these elements and at different points in their development – for example green growth is useful when the plant is establishing, but too much nitrogen when the plant should be flowering may result in too much lush growth at the expense of blooms.

OPEN POLLINATED

Seeds that are pollinated naturally by wind, insects, birds or humans.

ORGANIC MATTER

Organic materials that are naturally broken down over time into a form that provides nutrients and aerated structure to the soil.

OVERWINTER

The process of sowing plants such as annuals or perennials in autumn and either planting them out in their final position to establish over the cold months or keeping them in pots in a sheltered place until temperatures increase in the spring. Any planting out ahead of winter should be done while temperatures are still mild so plants have the opportunity to establish before the cold sets in.

PERENNIAL

A plant that will survive for several years. Some are more short lived or tender than others. Most can be propagated by root cuttings or division.

PERLITE

A lightweight medium derived from volcanic rock that can absorb several times its own weight in water. A useful material to add to potting compost to improve drainage in containers.

PHOTOSYNTHESIS

The biological process by which plants convert sunlight into sugars and carbohydrates.

PINCH OUT

To remove the top growth from a young plant to encourage it to regrow sturdier and bushier.

PLUG PLANT

A healthy, semi-established seedling. This usually refers to small plants for sale from commercial nurseries. Plugs come in different sizes, so check before you purchase.

POLLINATOR

A creature, usually an insect, that carries pollen from the male anther of a flower to the female stigma of the same or another flower. Pollinators are responsible for the reproduction of approximately 80% of the world's flowering crops. Bees, bats, moths, butterflies, hoverflies, wasps and beetles all carry out this vital work.

POT ON

To move a seedling or plant into a larger pot, providing more space for it to keep growing and prevent the roots from becoming stunted.

PRICK OUT

To remove a germinated seedling from a seed tray or module into a larger pot on its own.

PROPAGATE

To increase the number of plants by sowing from seed; taking root, stem or leaf cuttings; or by dividing existing plants.

PROTOZOA

Single-celled organisms present in the soil. They feed on bacteria, but also eat other protozoa, soluble organic matter, and sometimes fungi. Protozoa release excess nitrogen as they feed on bacteria and this can then be used by plants and other microorganisms.

REEL WIRE

Strong florist's wire used to make installations or wreaths, although wreaths can be made with strong twine as well.

SEPAL

The protective casing – usually green – of a flower bud such as a rose. Once these have reflexed back towards the stem a rose is ready to cut.

SEED LEAVES

The first set of leaves that appear when a plant germinates. These leaves look different to the true leaves that the plant generates next, which are a miniature version of the mature leaves.

TRUE LEAVES

The first set of leaves on a seedling that are identifiable as the plant you have sown. Orlaya's true leaves, for example, have a frondy appearance; the seed leaves that germinate first are long and thin.

VERMICULITE

A naturally occurring mineral with a neutral pH. It can absorb up to three or four times its weight in water, which it releases gradually. A layer of vermiculite on a seed tray helps to protect seeds against the fungus that causes damping off.

VOLUNTEER

A self-seeded plant that you can transplant to grow elsewhere.

WORMERY

A composting system that utilises red worms (*Eisenia fetida*) or similar species to process food scraps into vermicompost, a rich and fine soil substance that is extremely beneficial to soil and plant health.

INDEX

Note: page numbers in **bold** refer to information contained in captions.

RESOURCES

This list is by no means exhaustive. These are the books and suppliers that we have found useful and have come back to again and again.

BOOKS ON FLOWER GROWING

The Cut Flower Patch, Louise Curley, 2014, Frances Lincoln
Cut Flowers, Celestina Robertson, 2022, Frances Lincoln
The Cutting Garden, Sarah Raven, 2013, Frances Lincoln
The Flower Farmer's Year, Georgie Newbery, 2014, Green Books
Floret Farm's Cut Flower Garden, Erin Benzakein, Michelle M Waite, 2017, Chronicle Books
Floret Farm's Discovering Dahlias, Erin Benzakein, Jill Jorgensen, Julie Chai, 2021, Chronicle Books
Floret Farm's A Year in Flowers, Erin and Chris Benzakein, 2020, Chronicle Books
Grow and Gather: A Gardener's Guide to a Year of Cut Flowers, Grace Alexander, 2021, Quadrille Publishing

REFERENCE

Annuals and Biennials, Roger Phillips, Martyn Rix, 2002, Random House
The British Flowers Book, Claire Brown, 2018, available at: https://www.thebritishflowersbook.co.uk/
How Bad Are Bananas: The Carbon Footprint of Everything, Mike Berners-Lee, 2010, Profile Books
The Random House Book of Perennials: Volume 1, Roger Phillips, Martyn Rix, 1991, Random House
The Random House Book of Perennials: Volume 2, Roger Phillips, Martyn Rix, 1992, Random House
The Random House Book of Shrubs, Roger Phillips, Martyn Rix, 1989, Random House

INSPIRATION AND FLORAL DESIGN

Blooms: Contemporary Floral Design, 2019, Phaidon
Cultivated: The Elements of Floral Style, Christin Geall, 2020, Princeton Architectural Press
The Flower Fettler's Year, Sarah Statham, 2021, available at https://www.simplybyarrangement.co.uk/
From Seed to Bloom: A Year of Growing and Designing With Seasonal Flowers, Milli Proust, 2022, Quadrille Publishing
Garden Notebook, Constance Spry, 1940, J.M Dent & Sons
Hostess, Constance Spry, Rosemary Hume, 1961, J.M. Dent & Sons
How to Do the Flowers, Constance Spry, 1952, J.M. Dent & Sons
Simple Flowers: A Millionaire for a Few Pence, Constance Spry, 1957, J.M. Dent & Sons
Summer and Autumn Flowers, Constance Spry, 1951, J.M. Dent & Sons
Winter and Spring Flowers, Constance Spry, 1951, J.M. Dent & Sons
Vintage Flowers, Vic Brotherson, 2011, Kyle Books

SUSTAINABLE FLORISTRY

A Guide to Floral Mechanics, Sarah Diligent, William Mazuch, 2020, available at: https://agtfm.com/

THE GLOBAL CUT FLOWER TRADE

Gilding the Lily, Amy Stewart, 2009, Portobello Books
Holland Flowering: How the Dutch Flower Industry Conquered the World, Andrew Gebhardt, 2014, Amsterdam University Press

BOOKS ON DRYING FLOWERS

Cut and Dry: The Modern Guide to Dried Flowers From Growing to Styling, Carolyn Dunster, 2021 Laurence King Publishing
Everlasting: How to Grow, Harvest and Create with Dried Flowers, Bex Partridge, 2020, Hardie Grant

BOOKS ON GROWING IN OTHER AREAS - ALSO USEFUL TO FLOWER GROWING

Beth Chatto's Garden Notebook, Beth Chatto, 1988, Dent Ltd
The Five Minute Garden, Laetitia Maklouf, 2020, National Trust
For the Love of Soil: Strategies to Regenerate Our Food Production Systems, Nicole Masters, 2019, Printable Reality
How to Grow Your Dinner Without Leaving the House, Claire Ratinon, 2020, Laurence King Publishing
Organic Gardening the Natural No-Dig Way, Charles Dowding, 1988 Green Books
The Well-Tempered Garden, Christopher Lloyd, 1970, Gardeners Book Club

BOOKS ON BIODIVERSITY AND THE NATURAL WORLD

English Pastoral: An Inheritance, James Rebanks, 2020, Allen Lane
The Garden Jungle, Dave Goulson, 2020, Vintage
A Sting in the Tale, Dave Goulson, 2013, Jonathan Cape
Wilding, Isabella Tree, 2018, Picador
The Well Gardened Mind, Sue Stuart Smith, 2019, Harper Collins

SEED SUPPLIERS

Chiltern Seeds https://www.chilternseeds.co.uk/
Cotswold Seeds (great for ground cover crops) https://www.cotswoldseeds.com/
Green & Gorgeous https://www.greenandgorgeousflowers.co.uk/seeds/
Higgledy Garden https://higgledygarden.com/
Milli Proust https://www.milliproust.com/
Moles Seeds https://www.wholesale.molesseeds.co.uk/
Plants of Distinction https://www.plantsofdistinction.co.uk/
Seed Cooperative https://seedcooperative.org.uk/
Tamar Organics https://tamarorganics.co.uk/

COMPOST

Climate Compost https://www.thelandgardeners.com/soil-home
Dalefoot https://www.dalefootcomposts.co.uk/
Fertile Fibre https://www.fertilefibre.com/
Field Compost - Organic Soil Improver https://www.fieldcompost.co.uk/store/products/organic-soil-conditioner

DAHLIAS – REFERENCES, TUBERS AND ROOTED CUTTINGS

Halls of Heddon https://www.hallsofheddon.com/
Just Dahlias http://justdahlias.co.uk/
Peter Nyssen https://www.peternyssen.com/
Pheasant Acre Plants https://www.pheasantacreplants.co.uk/
Rose Cottage Plants https://www.rosecottageplants.co.uk/
Sarah Raven https://www.sarahraven.com/
Withypitts Dahlias https://www.withypitts-dahlias.co.uk/

BULBS

Jacques Amand https://jacquesamandintl.com/
Organic Bulbs https://www.organicbulbs.com/
Peter Nyssen https://www.peternyssen.com/
Sarah Raven https://www.sarahraven.com/

CHRYSANTHEMUMS

Chrysanthemums Direct https://www.chrysanthemumsdirect.co.uk/
Sarah Raven https://www.sarahraven.com/

BARE-ROOT ROSES

David Austin Roses https://www.davidaustinroses.co.uk/
Harkness Roses https://www.roses.co.uk/
Peter Beales Roses https://www.classicroses.co.uk/

PERENNIALS AND SPECIALIST NURSERIES

Always check for independent nurseries local to you.
Arvensis Perennials https://www.arvensisperennials.co.uk/
Beth Chatto https://www.bethchatto.co.uk/shop-plants.htm
Claire Austin https://claireaustin-hardyplants.co.uk/
Marshall's Malmaisons Email marshalls.malmaisons@btinternet.com or call 01473 822 400
Norfolk Herbs https://www.norfolkherbs.co.uk/
Woottens of Wenhaston for scented pelargoniums https://www.woottensplants.com/

ORGANISATIONS TO JOIN OR RESEARCH

Fairtrade International https://www.fairtrade.net/
Flowers from the Farm https://www.flowersfromthefarm.co.uk
The Gaia Foundation https://www.gaiafoundation.org/
Land in our Names https://landinournames.community/
Land Workers Alliance https://landworkersalliance.org.uk/
Organic Growers Alliance https://organicgrowersalliance.co.uk/
Soil Association https://www.soilassociation.org/

ADVANCED LEARNING ONLINE AND IN PERSON

The Business of Selling Flowers https://www.thebusinessofsellingflowers.com/
Floret online workshop https://www.floretflowers.com/
Forever Green Flower Company https://www.forevergreenflowerco.co.uk/
The Land Gardeners https://www.thelandgardeners.com/
Love 'n Fresh Flowers https://lovenfreshflowers.com/
Organic Blooms https://organicblooms.co.uk/

GROWERS OF WOLVES LANE

Growing is about abundance and productivity, but it is also about people and the importance of sharing skills and knowledge, commiserating and celebrating together. While some days we yearn for a four-acre rural plot, we would undoubtedly miss the camaraderie, advice and generosity of our fellow growers with whom we work cheek by jowl in this pocket of North London.

ACKNOWLEDGEMENTS

JOINT

To all at Wolves Lane - new and old - who we have shared this ramshackle, magical and frustrating site with. It is like no other place and will continue to evolve and grow, shaped by the wonderful people who take root there for a while. Particular thanks to all the good-natured growers who have allowed us to snoop around their beds and snip bolted veg and flowering cover crops, which we've coveted from the next glasshouse along.

To Iain Drury of Wolves Lane Allotments who tolerates our bindweedy, flowery chaos and always goes out of his way to help us.

To Emily Jones for the fireplace.

To Penny Snell and Shane Connolly for being so generous with their time and talking to us at length about their experiences of floristry and seasonality.

To Philippa Stewart for her kindness and patience and for always answering our dahlia-related questions.

To the RHS Lindley Library and their wonderful staff for archiving everything written by Constance Spry and for letting us read it all.

To Jo and Wendy of Organic Blooms who were one of our early inspirations in flower growing and show us all exactly how it should be done. And to all the members of Flowers from the Farm who share their knowledge and expertise so generously: let's keep helping each other to make seasonality the norm.

To our amazing WLFC team, most notably Carrie, Mandy and Pauline who support our growing operation even when we are chaotic and cranky. And to all the wonderful people who have volunteered their time with us over the last four years: there are too many of you to name without unforgivably forgetting someone but you are all part of our success and story and we are so thankful.

To Helen Lewis, Sophie Allen, Clare Double and Alice Kennedy-Owen of Pavilion Books, who have patiently nurtured and supported us on this multi-year voyage to write a book about some of the things we've learnt while growing flowers. Thanks for thinking we had something to say when we didn't and for helping us turn it into something we can be proud of.

To Charlie Ryrie, who has gently interrogated, suggested and helped corral our ramblings into a publishable state. Let's hope we can meet up and just chat about flowers next time!

To Aloha Shaw for absolutely and totally 'getting' what is beautiful about the shabby and messy site we work on and capturing it effortlessly in your photography. This book is beautiful because of you and your creativity.

MARIANNE

To David who has to put up with a lot of flowery debris in the house, quite a tense and snappy pregnant wife trying to write a book and who is never-endingly optimistic and supportive.

Jem, you make me so proud that you already love to forage, can identify a hollyhock and like coming to mummy's work to 'help'. You have since being in utero had to fit around said work and you do so with curiosity and a surprising amount of patience for a three-year-old. Always be a nature boy.

To Cass who has grown inside me during the writing of this book and has become more and more of a presence in the photographs as the year has gone on! I hope the love of flowers and plants has been seeded in you too.

To my parents. Without your early nurturing of my nature mind would I be doing what I am doing? Thank you for letting me run barefoot in the garden, for teaching me to be sharp eyed and able to spot the wonder of plants, nature and the countryside. You have always championed me, tolerated my emotional outbursts, heroically supported with childcare and encouraged me that there is something worth reading in what I have to say.

And to Camila. Soily fingered partner in crime, whose brain is more in tune with mine most of the time than anyone else's. Whose capacity to keep on keeping on is breathtaking and who is the person who always makes things happen. Every year of WLFC to date has felt like we've pushed ourselves to the edges of our comfort zone and sanity but I've loved undertaking this exhausting and exhilarating adventure with you. Let's not forget a moment of rest and celebration once this is all finished and before one of us suggests another hard and scary project!

CAMILA

To Effie and Michael Romain for being so supportive, especially with childcare when deadlines were looming and there were Covid-related closures.

To Raph, Marty and Clem for allowing me to hijack every family holiday with a garden visit.

To my mother for introducing me to the garden and

making me love it as much as she does and for always turning up with cake.

To *my* David who has shown unfathomable patience and empathy, for tolerating sleepless nights and all manner of book-related anxieties, while also being a first-time dad with a wife supposedly on maternity leave. Thank you for always supporting me in this labour of love.

To Bruno for accommodating a book in your first year of life. I hope we'll grow flowers together when you are old enough.

To Marianne for your ambition and resolve, for taking WLFC beyond just growing wonderful flowers, for thinking bigger than what needs to be cut or sown, potted on or planted out. For always thinking, tinkering, planning and plotting, that mind of yours that never stops. And for being the most patient and understanding friend and business partner, when I had a long list of chapters to write but also a tiny baby who didn't sleep. We got through it together but mainly with a lot of morning pep talks and encouragement from you.

'That is one of the things flowers do for you, they break down barriers and make for friendliness, they crumble that wall of shyness that stands between so many English men and women, sometimes making their first reaction to a stranger hostile rather than friendly; you might perhaps call it the freemasonry of flowers.'

Constance Spry, *How to Do the Flowers*, 1952

FLAT LAY INGREDIENTS

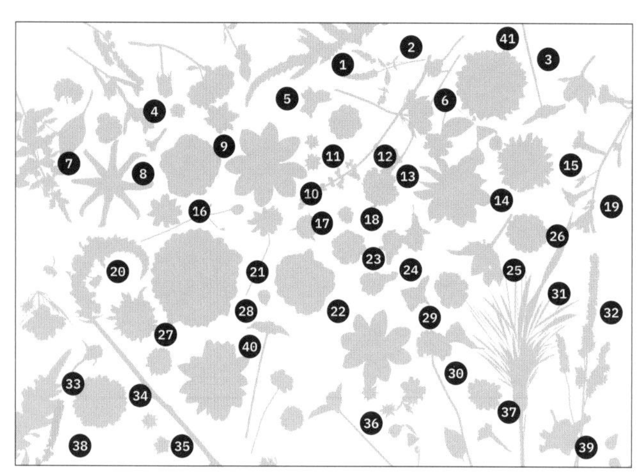

Autumn pages 28–29

Dahlias pages 62–63

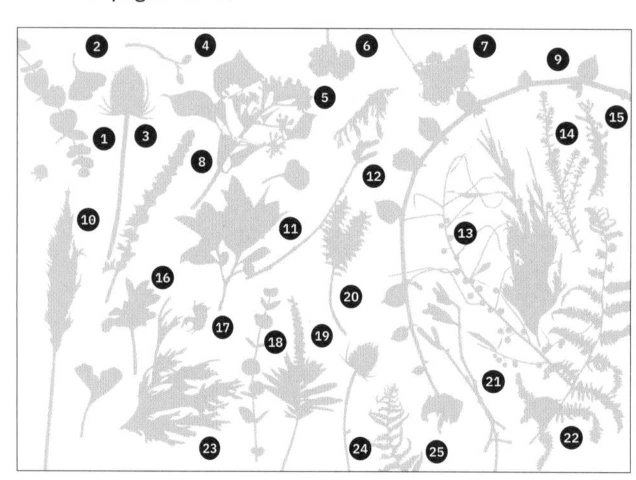

Winter pages 74–75

Autumn
1. 'Fox Tail' amaranthus
2. Mint
3. 'White' globe amaranth
4. 'Tinkerbell' tobacco flower (*Nicotiana alata*)
5. 'Potomac Lavender' snapdragon
6. 'Purity' cosmos
7. Mint
8. *Dahlia* 'Honka Fragile'
9. 'Lady of Shalott' rose
10. *Dahlia* 'Totally Tangerine'
11. Borage
12. *Linaria purpurea* 'Canon Went'
13. 'Snowmaiden' scabious
14. *Dahlia* 'Waltzing Mathilda'
15. *Zinnia* 'Señora'
16. 'Polidor Mixed' cosmos
17. 'Orange' strawflower
18. *Zinnia* 'Jazzy Mix'
19. *Nicotiana* 'Whisper Mixed' (tobacco plant)
20. *Verbascum* x *hybrida* 'Southern Charm'
21. *Dahlia* 'Otto's Thrill'
22. 'Lichfield Angel' rose
23. 'Popsocks' cosmos
24. 'Potomac Lavender' snapdragon
25. Acidanthera
26. 'Queeny Lime Orange' zinnia
27. 'Tower Chamois' China aster
28. Love in a puff
29. Dahlia grown from open-pollinated seed
30. 'Merlot Red' scabious
31. 'Sprinkles' ornamental grass
32. 'Pink Pokers' statice
33. Showy amaranth 'Hot Biscuits'
34. Chrysanthemum 'Talbot Parade Salmon'
35. Borage
36. 'White' globe amaranth
37. *Rudbeckia hirta* 'Goldilocks'
38. Velvet trumpet flower
39. 'Bright Rose' strawflower
40. *Dahlia* 'Cameo'
41. *Dahlia* 'Salmon Runner'

Dahlias
1. *Dahlia* 'Salmon Runner'
2. *D.* 'Honka Fragile'
3. *D.* 'Leila Savannah Rose'
4. *D.* 'Arabian Night'
5. *D.* 'Twynings White Chocolate'
6. *D.* 'Apricot Desire'
7. *D.* 'Pam Howden'
8. *D.* 'Crème de Cognac'
9. *D.* 'Penhill Watermelon'
10. *D.* 'Apache'
11. *D.* 'Otto's Thrill'
12. *D.* 'Rip City'
13. *D.* 'Cornel Brons'
14. *D.* 'Perch Hill'
15. *D.* 'Senior's Hope'
16. *D.* 'American Dawn'
17. *D.* 'Wizard of Oz'
18. *D.* 'Brown Sugar'
19. *D.* 'Polventon Kristobel'
20. *D.* 'Wine-eyed Jill'
21. *D.* 'Cafe au Lait'
22. *D.* 'Verrone's Obsidian'
23. *D.* 'Franz Kafka'
24. *D.* 'Schipper's Bronze'
25. *D.* 'Darkarin'
26. *D.* 'White Ballerina'
27. *D.* 'April Heather'

Winter
1. Eucalyptus
2. Gingko
3. Teasel
4. Rosehips
5. Ivy
6. Sedum
7. Old man's beard (*Clematis vitalba*)
8. Buddleia
9. Physalis
10. Common reed (*Phragmites australis*)
11. Strawberry tree

12. Catkin (hazel)
13. Asparagus
14. Sea rosemary
15. Curry plant
16. Euonymus
17. Old man's beard
18. Eucalyptus
19. Crimson bottlebrush
20. Hebe
21. Catkin
22. Bracken
23. Conifer
24. Teasel
25. Bracken

Dried flowers

1. Feverfew
2. Sea rosemary
3. Larkspur
4. Lady's mantle (*Alchemilla mollis*)
5. Tagetes
6. Tansy
7. China aster
8. Broadleaf plantain
9. Sanguisorba
10. Artemisia
11. Chinese forget-me-not (*Cynoglossum amabile*)
12. Globe amaranth
13. Quaking grass

Spring

1. Flowering currant
2. Spurge (euphorbia)
3. *Tulipa sylvestris*
4. 'Tête-à-tête' daffodil
5. *Calendula officinalis* 'Snow Princess'
6. Forget-me-not
7. *Anemone* 'Galilee Pastel Mix'
8. *Fritillaria uva-vulpis*
9. 'Vulcan' sweet William
10. Dwarf comfrey
11. 'The Bride' anemone
12. 'Cloth of Gold' wallflower
13. Greengage blossom
14. *Tulipa turkestanica*
15. 'Vulcan' wallflower
16. *Anemone mistral bianco centro nero*
17. Willow
18. Broom
19. Grape hyacinth
20. Heart's ease (*Viola tricolor*)
21. 'Vulcan' wallflower
22. Cherry blossom
23. Heather
24. *Calendula officinalis* 'Touch of Red'
25. Dwarf comfrey
26. Fumitory
27. Rocket
28. Periwinkle
29. *Anemone coronaria* 'De Caen'
30. Grape hyacinth
31. Hellebore

Summer

1. Sorrel
2. Aquilegia
3. *Cynoglossum amabile* 'Firmament'
4. 'Carmine King' larkspur
5. 'Wiltshire Ripple' sweet pea
6. 'Lichfield Angel' rose
7. Honesty
8. *Achillea millefolium* 'Summer Pastels'
9. 'Thai Silk' California poppy
10. Catmint (Nepeta)
11. 'Lady of Shalott' rose
12. Argentine forget-me-not
13. Columbine (*Aquilegia vulgaris* 'Nora Barlow')
14. Coriander
15. Wild carrot
16. Sicilian honey garlic
17. Catmint (Nepeta)
18. Corncockle
19. 'Ivory Castle' California poppy
20. *Calendula officinalis* 'Cantaloupe'
21. *Cynoglossum amabile* 'Firmament'
22. Geum 'Mai Tai'
23. *Argentine* forget-me-not
24. Nigella
25. Orlaya
26. *Achillea millefolium* 'Summer Pastels'
27. Chives
28. 'Sooty' sweet William
29. Sainfoin
30. Astrantia

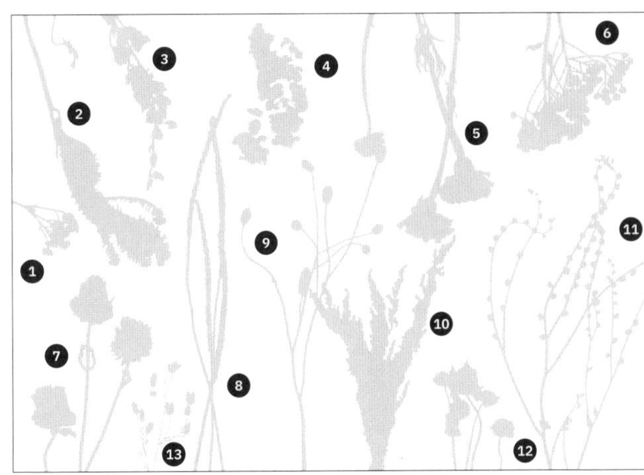

Dried flowers pages 100–101

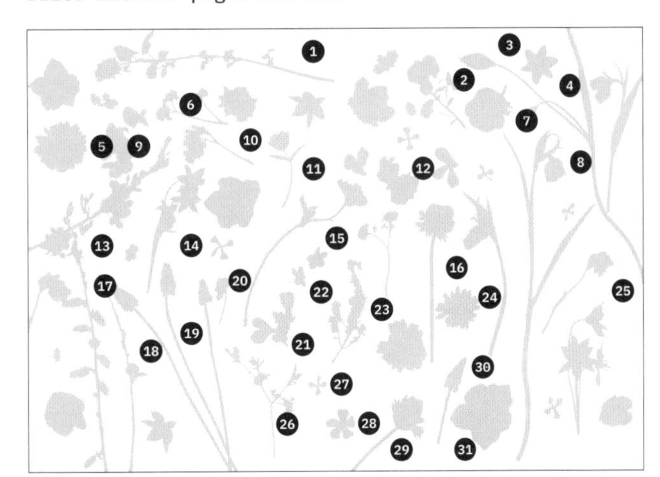

Spring pages 110–112

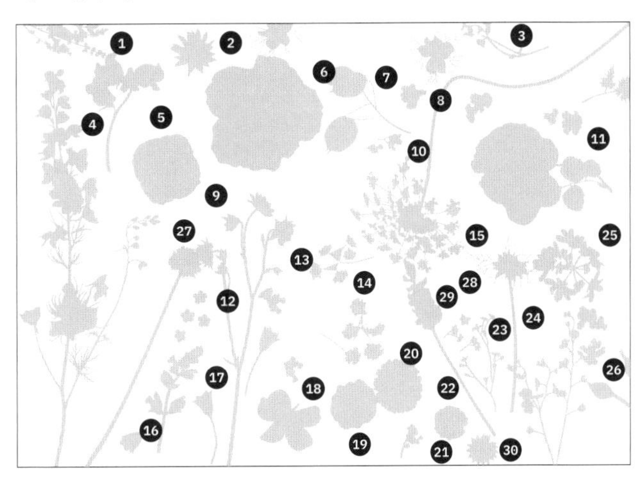

Summer pages 150–151

FLOWER VARIETIES

DAHLIA STAIRCASE (PAGE 64) INCLUDES:
Amaranth 'Hot Biscuits'
Cosmos 'Cupcake Blush'
C. 'Popsocks'
Dahlia 'American Dawn'
D. 'Bishop of York'
D. 'Brown Sugar'
D. 'Darkarin'
D. 'Honka Fragile'
D. 'Jowey Winnie'
D. 'Otto's Thrill'
D. 'Perch Hill'
D. 'Salmon Runner'
D. 'Waltzing Mathilda'
Nasturtium 'Ladybird Rose'
Nicotiana 'Whisper Mix'
Orach
Rudbeckia 'Goldilocks'
Sunflower 'Magic Roundabout'
Wild fennel
Wisteria foliage
Zinnia 'Jazzy Mixture'
Z. 'Senora'

DRIED-FLOWER NEST (PAGE 102) INCLUDES:
Alchemilla mollis
Ammi majus
Bamboo
Bunny tail
Cardoon
China aster
Cornflower 'Black Ball'
Dock
Feverfew
Honesty
Hydrangea
Larkspur
Love in a puff (at the bottom)
Lupin seed head
Nipplewort
Ranunculus
Scabious
Sicilian garlic
Statice
Strawflower
Tansy

WILD MANTELPIECE (PAGE 142) INCLUDES:
Anemone 'The Bride'
'Belle Epoque' tulip
'Copper Image' tulip
Jasmine
Narcissus 'Cheerfulness'
N. 'Thalia'
Pearl bush
Pieris japonica
Rocket flower
White cherry blossom

TABLE RUNNER (PAGE 188) INCLUDES:
'Alan Titchmarsh' rose
Nepeta 'Six Hills Giant'
Chamomile
Corncockle
Cornflower
Dock
Galega
Nigella
Nipplewort
Quaking grass
Sanguisorba
Wild carrot
Wild grasses

Dahlia Staircase

Dried-flower Nest

Wild Mantelpiece

Table Runner